C. S. Coy grew up in a small town near Western New York. Majoring in Health Information Technology in college, his true passion is writing, beginning when he was a young kid. With numerous ideas and short stories in his head, it wasn't until his college years that he began to pursue his dream to be a writer. He hopes to share his love of writing with people who can find hope and inspiration through his stories.

Dedication

In dedication to T. B.

C.S. Coy

THE UNEXPECTED

AUSTIN MACAULEY PUBLISHERS™
London * Cambridge * New York * Sharjah

Ordering Information:
Quantity sales: special discounts are available on quantity purchases by corporations, associations, and others. For details, contact the publisher at the address below.

Publisher's Cataloguing-in-Publication data
Coy, C.S.
The Unexpected

ISBN 9781641823050 (Paperback)
ISBN 9781641823067 (Hardback)
ISBN 9781641823074 (E-Book)

The main category of the book —. Science fiction

www.austinmacauley.com

First Published (2018)
Austin Macauley Publishers Ltd™
40 Wall Street, 28th Floor
New York, NY 10005
USA

mail-usa@austinmacauley.com
+1 (646) 5125767

CAUTION: THIS STORY CONTAINS HEAVY SCENES THAT INVOLVE DEPRESSION AND SUICIDE. THE MISSION OF THIS STORY IS TO RAISE AWARENESS AND TO SUPPORT OTHERS IN NEED. IF YOU, OR ANYONE YOU KNOW, ARE IN NEED OF HELP, PLEASE DO NOT HESITATE TO TALK. WE ARE ALL HERE FOR YOU AND WE SUPPORT YOU. THANK YOU.

- C. S. COY

Prologue

The year is 2025.
Up above, a vast scene of grey clouds moved across the sky as a calm fall breeze could be felt in the air. From afar, the city of Hollandview stood. All around, windows were shattered with debris sitting in the streets. As quiet as it could be, no sign of life showed within this lonely city.

Three years have passed since the invasion of the creatures known as Instinctive Outside Beings (IOBs). Wiping out cities and towns, the creatures wrecked their havoc on the world while taking the lives of millions. Fighting back, the world tried to counter this new threat while they were ultimately destroyed. With militaries and countries being obliterated, the world slowly began to die while the creatures only continued to live.

Though it all seemed to be lost in the world, it wasn't until south of the city limits of Hollandview where peaks of life could be seen in a section called Sector A. With buildings still left standing in the area and roads cleared of gravel, people dressed in rags still walked amongst each other as they talked and lived their days. With their faces filled with dirt, they all moved around the streets of the city, as some rested eating scraps of food while they watched the scene of people moving around. Even above in buildings, children and adults sat in the broken window sills as they looked up into the sky hoping sunlight would come out some time.

While most of the people resided in the streets of the sector, soldiers who guarded the city called IOB Containment Forces or IOBCFs, stood above on the building tops, looking at one another as they carried their weapons. Waiting and watching for anything to come.

Wearing a navy blue jacket uniform and a black vest with the logo of a fist and a ribbon, these containment forces were the only defenses against any threats to Sector A as they only consisted of 800 soldiers, put in charge of defending over the 4,000 civilians that reside in the city.

Outside Sector A, nothing but abandoned highways could be seen. Full of abandon cars and garbage, a clear scene of woodland terrain covered the outside of the city. Going on for miles, nothing could be seen. No town, no cities. No roads.

It was there though, where a man named Clay could be seen.

Part 1

Chapter 1

Clay was standing in the woods as he could feel a warm breeze of air pass around him. Standing with his back turned, he wore a black Velcro stripped t-shirt and black pocketed pants, as he wore boots to top it all off.

Around him, he wore a metallic orange armor that wrapped around his upper body and chest. On his forearms, he had a bulky smooth guarding down to his wrists as he wiggled his fingers surrounded in the armor.

Before him, he stood looking into a puddle, he could see his reflection within it. Seeing his short black hair and the light dirt trails across his cheeks and chin, he simply closed his eyes as the thin tall trees surrounded him.

Standing in place, the gloomy scene surrounded him as the grey clouds above began to release a flurry of ash down on him.

With the ash falling lightly, he soon began to have images appear within his head as he began to relive old memories.

Looking, he found himself in the city. His eyes wide open while his body shook. The air felt hot as ashes fell down to him. Around, he could see the blur of people's faces around him.

He soon looked up. The scene of war over his head.

Planes and jets rocketed over the city. The ground shook as explosions rocked the very earth. Buildings tumbled down while people could be seen running.

Around him, strange ships and crafts attacked the city.

Furiously, they shot down jets and planes. They attacked and obliterated people below while they crushed anything that stood in their way.

With the fire and smoke crowding into Clay's lungs, he stood nearly choking as he knelt down covering his face. Looking around, he furiously tried to see what was before him. Struggling to stand, he soon looked before him and saw his wife Mya lying motionless on the ground until…

Opening his eyes, he began breathing heavily as he fell to one knee beside the puddle before him. Breathing heavily, he gasped for air as he grabbed his chest as he tried to calm down. Seeing himself in his reflection, he saw a different man than the one standing as he looked into someone's face that was scared and helpless.

Seeing it, he eventually closed his eyes gripping the dirt under his hand as eventually he began clearing the thoughts in his head.

Eventually getting himself to relax, he soon began to calm down when all of a sudden, his ear piece began to ring in his pocket. Standing up then, he slowly took a deep breath as he took out the piece and began to talk.

"This is Clay," he began in a tired yet calm voice.

"Clay, this is Mya," a voice nervously said as he quickly recognized it. "Mya…are you alright?" he asked as she spoke.

"I'm fine, but I just received a distress call coming from the city" she said rapidly as he thought. "Can you transfer it over?" he then asked.

Quickly transferring over the call, Clay listened carefully as the recording seemed distorted.

"This is Lt. Garrett reporting in from Hollandview city…IOBs have infiltrated the defense lines…we are requesting immediate back up as quickly as possible…" she said as Clay heard gun shots and yelling in the background. "Again if anyone can hear me, we need immediate back up now!" she said as Clay looked ahead hearing a loud crack echo down to him in the woods. Hearing the sound pass around him, he quickly headed towards it.

Running towards the sound, he eventually worked his way through the woods and found himself standing over the edge of a cliff. Immediately stopping, he soon looked ahead as he could see the city from afar. Looking toward it, he could see a black trail of smoke beginning to rise up into the sky. Noticing it, he soon pressed on his earpiece as he spoke to Mya.

"Mya, did you hear that?" he then asked.

"I did…but what's going on out there?" she asked as Clay shook his head.

"I don't know," he began. "But something's not right. I'm going in to help!" he said looking at the city and stepping back.

Reaching into his belt as he pulled out a small disk, he soon threw it out before him. Expanding open in the air, it eventually turned into a circular board that hovered off the ground.

Stepping forward onto it, his feet were soon latched on as it slowly began hovering over the edge of the cliff. In the air, he looked ahead to the smoke continuing to rise from afar while he soon jetted off in the direction of the city that was under attack…

While he headed towards the city, the people in Sector A began to head out towards the south evacuation points. All leading to the abandon highways from behind, they desperately ran in panic to seek sanction within the woods.

Ahead, IOB Containment Forces scrambled through the Sector. Heading to the defenses up North near the sector entrance, dust and debris lay in the streets. Fire lit up the sidewalks as casings of bullets slowly descended down to the ground.

There, the IOB Creatures could be seen.

Hundreds of them crawling and running into the city section. With grey, rubbery, wrinkled-up skin, they towered over the soldiers. They had a blade-like hand on one side and sharp claws on the other. Their teeth barred before their rounded massive heads as a blank face hung before them. With their massive eyes moving on the side of their heads, the creatures attacked tearing into the soldiers in their way.

Firing, the IOB containment forces tried to hold off the wave of creatures.

The creatures tore into any soldier and struck them down beneath them. And when a soldier went down, the creatures would only turn to the next as if nothing would stop them.

Being overwhelmed, soldiers began running away as they went to join up with the rest of the evacuation. But as they did, a woman among them raised up her gun with one hand on the trigger and the other on the barrel. Firing, she wore a containment force uniform and helmet as she ran past the soldiers stepping back.

Running to the front lines, she stopped, looking at the creatures a few feet away from her.

Firing towards them as creatures fell back, she quickly pulled out her radio and spoke as the soldiers rallied around her.

"Nick, this is Lt. Garrett. Do you read?" she asked as from afar in the city leading people out, a man named Nick Coast who had short dark-brown hair and the commander of the IOB Containment Forces, directed the people out of the city as he spoke to the Lt.

"This is Nick, how's it looking?" he asked, as she quickly lifted her gun and shot back an IOB while she spoke.

"We're getting owned out here, Nick! Where the hell are you?! We need some aid out here now!" she said as the Nick shook his head.

"We're trying to get these people to safety first! If we don't get them out, then we'll be in trouble," he said as Emily shook her head.

"If we don't hold these things off now, we won't have anything left to protect! I need your soldiers now!" she said as Nick shook his head.

"Just hang in there a bit longer! You need to hang on!" he said as Emily slammed her radio on the ground. "There is no time!" she then said as all of a sudden she looked ahead of her, as the IOBs creatures looked to her and soldiers. Standing still, their mouths began to glow red. Loading an aura-like ball in their mouths, the Lt. soon watched as soldiers beside her began to run away while the creatures shot out towards them. Quickly ducking down, she fell to the ground as the shots flew over her.

Behind her, the soldiers retreating were hit with the blasts.

Falling hard, the soldiers fell while the blasts burned into their skin or blasted off their limbs. Bodies filled the streets as the blasts shot all around.

Breathing heavily, she soon watched as an IOB screeched towards her. Seeing it coming, she quickly lifted a knife beside her. As it ran into her she stuck the blade into it. Not killing it, she lost grip of the blade as it was still stuck in the beast which held her down with its massive claws around her neck. Choking her as she lay on the ground struggling, she could only watch the drooling teeth hang over her, as it raised its bladed hand back preparing to strike her.

Seeing herself about to be struck, she then watched as before her, as a blade pierced through the IOB's body. Sticking out towards her, the IOB moaned while Clay stood behind it. Pulling it off her, he then swung the creature around before him as he threw it out on the ground, before the other IOBs.

Sitting up slowly, the Lt. looked at Clay as he had two long blades hanging out of the armor on his arms. She could only wonder who he was, as he soon glanced over his shoulder and spoke to her.

"You need to go…now!" he began quickly as she slowly shook.

"No, I can't. I have to fight!" she said as the creatures all began screeching and roaring amongst each other, while Clay shook his head slowly.

"There's no time…get to the other soldiers and bring them back here," he said.

She quickly spoke, "But what about you?"

Clay had anticipated her question. "I'll hold them off until then," he said, looking at her. "Just go!" he said as she eventually felt herself nod and get up. As she began walking away, she stopped and turned to Clay and looked at him for a moment, saying, "Good luck!" while she began running from the scene.

Giving her time to escape, he soon looked back at the IOBs.

Standing with his blades held beside him, an IOB ahead soon held up its blade, as it let out a terrifying screech into the air. Looking at them, the IOBs all quickly began to move towards him. He watched the herd of creatures tumbling around the streets.

Some running on two legs as they swung their arms back and forth, others dropped to all fours while they ran ahead tearing into the concrete.

Breathing slowly, Clay simply looked at the creatures before him as he quickly lifted up his blades. Crossing them, he stood clinching his fists together as the creatures ran closer. Taking a deep breathe, he soon looked to the first IOB before him.

Lunging, Clay quickly spun. Gliding the blades across its chest, the IOB creature quickly dropped to the ground as it slid past him.

Standing back up, he looked before him as an IOB threw its blade over him. Ducking down, he quickly threw the blade inwards, as he pulled it out.

With the creature falling, another IOB ran to him slashing its claws at him. Raising up his arm as its claws scraped across his armor, he quickly spun, striking the creature down. Screeching as it fell, Clay looked ahead of him. Herds of IOBs continued to run towards him. Throwing his arms down, he quickly then ran towards them.

He ran to one piercing a blade through it as he spun around it and cut down another. Taking down as many IOBs as he could, he quickly fought each IOB while he guarded himself from their vicious attacks.

Striking each one down, it wasn't long before he began to struggle to hold off the hundreds of IOBs before him.

Getting overwhelmed, he reached beside him as he pulled out a device that went off on the ground in front of him. Releasing a series of flashes that blinded the creatures around him, Clay quickly went up to one, piercing a blade through its head as he pulled it out and grabbed another device.

Smashing it on the ground, a mist of white smoke began to fill up the streets around him.

With the creatures unable to see, Clay began using the smoke to disguise himself while he quickly ran to the IOBs. Moving his blades forward, the IOBs tried countering as Clay moved all around them.

While he did, the IOBs outside the smoke looked on as they simply dropped down and fired blasts into the scene. Crashing down around him, Clay moved around while he watched the blasts shoot down into the street or hit other IOBs.

Kneeling down waiting, he soon watched as an IOB ran up beside him as he quickly stood up. As it crashed into him, Clay fell back while the IOB hung over him. Screeching, it threw its bladed hand down to the ground while it caught the side of Clay's stomach. Gliding across him while the blade smashed into the concrete, Clay quickly kicked back the creature. Stepping back, it grunted and moved around while he stood back up.

Looking at the creature, it soon screeched back towards him. He stood up and then cut off its head as it rolled to the ground. With its body falling down, Clay then tried to catch his breath while he stood. Feeling pain, he looked to the blood on his side. Putting his hand over it, he looked back beside him while the smoke cleared.

He could see the herds of IOBs around him. Snarling and screeching they all stood looking at him. With so many around him, he began to think about what to do. Standing while he thought, an IOB ran up to him as he raised up a blade. With its jaws open, he held the creature back while he held the blade in its mouth. Holding it back, he pierced the blade through it and threw the creature to the side.

Stumbling back, an IOB screeched behind him while he quickly turned, gliding the blade across it.

Falling to one knee then, he felt himself slowing down. Quickly an IOB went up to him, slashing its blade across his arm.

Turning, he cut off the creature's arm. He then pierced a blade through it.

Struggling to stand then, he soon watched as an IOB let out a blast before him. Shooting out towards him, he fell back as the blast began to burn the flesh on his arm. On the concrete, he laid his head back as he tried putting his hand over the burn. Grunting in pain, he could only look at the vast army of IOBs still before him as they crept slowly towards him.

Trying to force himself up then, Clay clinched his teeth together tightly, trying to stand as his body began to wear out. With pain overwhelming him, he realized that he couldn't get up as he tried to fight.

Seeing IOBs running towards him with their bladed hands up in the air, he could only watch silently as they went up to him, about to strike when all of a sudden, a piercing noise soon went off in the sky.

Hearing the noise, Clay began to cringe at the noise being set off. Keeping one eye open, he looked at the IOBs before him. He soon looked up to see a black metallic IOB ship hovering in the sky.

Seeing the small ship, Clay looked at the IOBs all before him, as they soon lowered their blades and began muttering to each other.

The creatures then looked back at Clay as they simply snarled to him. Not knowing what they were doing, Clay could only watch as the creatures all soon turned away and began retreating from the scene.

With the IOBs heading out, he merely could only shake his head in confusion as he looked back up to the IOB ship above him that soon began hovering away in the direction of where the IOBs ran.

Leaving the scene, it wasn't long before Clay eventually made it back to his feet. Standing while he looked out from Sector A, he watched the IOB creatures with their ship, retreat away from the city. He was left only to wonder to himself.

Chapter 2

With the IOBs retreating out of Sector A, Clay soon headed back. Walking through the streets of the sector, he held his hand over his wound while he looked around to see soldiers running around him. Seeing them go, he turned around and could see Nick. Nodding to him, they began walking through the streets together as they spoke.

"Man, am I glad to see you," Nick then began. "It's been crazy out here. People are all shaken up by this. If you didn't make it here when you did, I don't think we would have made it," he said as Clay shook his head while he spoke.

"Yeah well, I wish I would have gotten here a little faster though…" he said, his sentence trailing off into thoughts. "What happened here?" he asked as Nick shook.

"I wish I could tell you…It all happened so fast. An explosion occurred north of the sector's entrance, next thing I know, IOBs began pouring into the streets." he said as Clay looked down while he spoke.

"I've never seen so many IOBs together at once before…" he said looking down. "Something must have led them here."

"You think so?" Nick asked as Clay nodded.

"IOBs don't usually attack in herds…at least not like this," he said as Nick shook. "What do you think they were planning?" he asked as Clay spoke.

"I'm not sure," he said to him. "But one thing I know for sure is something doesn't seem right about all this" he said as Nick nodded.

"Yeah maybe," he began as he thought. "…it doesn't help that the fact we were caught off guard today either…we didn't even get to the alarm system in time," Nick stated as they stopped while Clay looked around and spoke.

"What do you think happened there?" he asked as he looked back to Nick.

"I have no clue…but regardless, it's our job to warn the people. We'll have to try to put a team together to investigate what really happened. See if we can try to prevent it from happening again," he said as Clay agreed.

"Well…keep me posted if you find anything," he said as Nick nodded.

The two were looking around while they spoke, Clay soon watched as a soldier beside them set down a body wrapped in a bag. Watching the solider move away, he then looked all around as more bodies were being lined up across the sidewalk. Seeing the bodies there, Clay could only shake his head quietly while he stood. Feeling emotions overwhelming him, he soon looked to Nick and spoke.

"How many?" he asked as Nick slowly spoke.

"Too many to count right now," he said. "…there's many still missing too. But by the looks of it, it seems only a few civilians were caught in the action. Mostly soldiers hurt in the blast," he said as Clay shook thinking as Nick padded him on the shoulder.

"Hopefully, we can do better next time…In the meantime, I better get back to work. We have a lot to do here" he said as Clay nodded. "Thanks again, man…we owe you one," he said as he soon left to join his soldiers ahead. As he did, Clay then looked back to the bodies lying in the street before him. Looking at them, he couldn't help but blame himself for being too late to the scene and letting the IOBs in. With guilt overwhelming him, he could only close his eyes and turn away while he continued to work his way up the street.

As he did, Emily stood looking at him when Nick appeared beside her, while she spoke. "Hey, you actually know him, Nick?" she began as Nick looked to Clay and nodded.

"You could say that. We go back. He's a good friend," he said as he walked away while Emily merely stood looking at Clay leaving.

A few moments later, Clay worked his way down the street as he went into a tent set up next to the soldier's supply depot. With soldiers running all around outside, a woman named Liz stood beside Clay as he sat up on a table while she looked to his wounds.

With long red hair, she wore a pair of blue worn scrubs and a grey zipper jacket as she was the same age as Clay. Helping him, she soon gave him pills to swallow as she spoke while he took it.

"Well, I have to say, you're lucky Clay…the cuts on your body don't look deep enough to put you in any immediate danger…but I can tell you're in pain," she said lightly as Clay nodded sitting on a table. "The pills I gave you should take the edge off a bit…But are you sure you don't want any stitches?" she asked, looking at him as he shook.

"No, it's alright Liz…I'll do that myself but thanks," he said nodding to her. Putting his shirt back down, Liz could only shake her head while she spoke to him.

"You know…it amazes me to see you still doing what you're doing, Clay," she began slowly as Clay looked to her. "I mean we're twenty-eight now and…I guess I'm just surprised to see you still doing this," she stated while he shook.

"Well…I guess someone has to do it," he said as he sat.

"True, but don't you ever get tired of it? Tired of all the fighting?" she asked as Clay sat thinking to himself for a moment as he nodded.

"At times…but no matter what I do, I can't stop fighting…otherwise that would mean giving up…and you know I can't do that," he said as she looked to him when all of a sudden they both looked up to see a woman walk in quickly with long straight brown hair as she was in a IOBCF uniform.

"Liz," she began as Liz turned to her.

"Oh my god…Holly," she quickly began running up to her and hugging. "I thought you might have got caught in the blast," she said as the soldier, Holly shook.

"I know…I should have come sooner but I had to help with the evacuation," she said holding her while Clay looked to them. Letting out a smile as he looked to them, he slowly got off the table and grabbed his armor. Attaching it back to his chest, Liz looked at him and soon spoke while he stood.

"Oh um Holly…this is a good friend of mine, Clay," she said as Holly looked to him while he walked up to her.

"It's nice to meet you," he began shaking her hand while she chuckled and looked to Liz and spoke.

"Oh my god…haha…You're…You're him, aren't you?" she began as she looked to Liz and shook. "You didn't tell me you knew him!" she said as Liz shook her head.

"I guess it slipped my mind," she said as Clay smiled a bit as he looked to Holly.

"Liz and I are old friends. She's helped me out a lot," he said as Holly shook as Clay then looked to Liz and spoke. "Well I better get going…it was good to see you again Liz…And I'm glad you've found someone finally," he said as Liz shook as she looked to Holly and quickly spoke.

"What? Oh no…no…no, me and Holly are just friends," she said as Clay nodded as he spoke.

"I know what it's like to love someone Liz," he began slowly as she looked to him. "You don't have to hide that with me," he said as he nodded to both of them and walked away from tent. While he did, Liz couldn't help but smile as she slowly held Holly's hand beside her.

Walking out of the tent then, Clay looked around as more soldiers ran around him. Stepping ahead as he pulled out his disk, he soon stopped as someone called out to him.

"Hey wait!" a voice began as Clay looked back to see Lt. Garrett walking up to him. With her blonde hair put back, she soon held her helmet beside her as she began to speak.

"I didn't get the chance to thank you back there" she said as Clay stood. "I don't usually have people saving me much…Name's Lt. Emily Garrett, by the way. I'm new here." she said as Clay nodded.

"I can tell you've faced IOBs before…" he said as she nodded.

"You could say," she stated. "I've dealt with a lot of IOBs outside of the city before I came here…I learned the hard way I guess…" she stated her thoughts. "…It's weird though, how quickly soldiers turn away from you when they see an IOB…that's what happens when you have a bunch of 18 and 19 year olds defend the city," she said as Clay looked around while he thought.

"Well, do you blame them?" he said looking back to her. "It's why we're here…it's why they need us," he said as she shook.

"Maybe. But I don't care much about what people are like or what they do with their own time. They should survive on their own and defend themselves if you ask me," she said as Clay nodded.

"Well, we have two different opinions then," Clay then said turning away as she smiled.

"I apologize if I seem a bit rough around the edges. It's always the way I've been," she then began. "I do care about this city and the people in it…It's just that I learned to survive by fighting…whether it's in my blood or fighting IOBs, it fills a void within me, I guess" she said as Clay stood. "Well, anyways, I take it you must be the guy that everyone talks about. ""Looks like the rumors are true. These people really look up to you," she said as Clay shook.

"Maybe…but they shouldn't. I've done nothing for them," he said as she shook. "Defending the city against IOBs…is that considered nothing?" she asked as he spoke.

"Not when people are dying…" he then said grunting in pain as she looked to him put his hand over his wound.

"You going to be ok?" she eventually asked as Clay shook.

"Yeah…I'm fine," he then said. "I just need to get back," he said throwing out the disk before him as it deployed open. He slowly stepped onto it. Hovering in place in the air, Emily looked to him as she walked up to beside him as she couldn't help speak.

"So you got a name?" she asked while Clay stood on his board looking at her. "It's Clay…Clay Treston" he said as she soon spoke.

"Well Clay…it was nice to meet you," she then said as he nodded his head and looked away while he hovered away from the scene, to head back home.

Chapter 3

A few moments passed by. Clay found himself standing still on the abandoned highway roads. Holding his head down while he held one foot on his board, he slowly looked up as he could see people walking pass him. Returning back to the city sector, he could see the looks on their faces. The looks of fear and pain. Covered in debris or blood, the people were tired. Tired of the pain and suffering they've had to endure over the past three years. The amount of losses they've had to deal with while trying to keep themselves carrying on. The temptation to believe that tomorrow would be better. With them all experiencing this at once, he knew only more pain lay ahead as he could only look away and continue home.

Eventually hovering back into the woods, he nimbly walked down to a trail as he slowly then stepped down. His board formed back into a disk, he slowly picked it up and clipped it back to his belt.

Soon, working his way through the woods, he found himself walking out into a clear field. With the grass turning yellow around him, he continued following a thin trail that was leading him to an old cabin.

The woods faded around the cabin as if it had been there for some time. A small porch could be seen by the cabin as the door opened and Mya stepped out as she quickly looked out to see Clay.

Wearing a long black sleeved sweater and jeans and with her short black hair, she quickly walked away from the porch, and then ran to him as he stopped and fell to one knee.

With her beside him, he slowly shook his head while he looked up to her and spoke.

"I'm…I'm sorry…I took so long Mya," he said as she had a tear run down from her face as she shook her head.

"No…it's ok," she said smiling. "I'm just glad you're ok," she said as he shook.

"Well…almost," he said lifting his hand off the side of his stomach. Looking at the wound bleeding out, she quickly helped him up as they slowly headed for the cabin.

Slowly opening the door as it creaked open, the outside light shone through the cabin as light dust could be seen floating in the gloomy room.

Helping him walk in slowly, they headed straight to a small round table beside the cabin window as Mya pulled out a chair.

"Here, sit," she said as she helped him sit. "I'll get some wet towels and a kit ready," she began as she then closed the door and turned beside her to the kitchen area of the cabin where she began opening the cabinets filled with canned food and supplies.

Getting the supplies ready, she eventually sat calmly in front of Clay while she began cleaning out his wounds. Shaking her head while she worked on him, she soon looked to him and slowly spoke.

"Clay, what…what happened down there?" she then began as Clay shook.

"The IOBs they…they came into the city…they…the IOBCF…they couldn't stop them…there were just too many," he said as he shook. "I tried to get down there as fast as I could…but it was too late…they were already there," he said as she spoke.

"I…I don't understand," she then began. "It's been three months since an IOB has been spotted in the city…why? Why would they come now?" she asked as Clay shook.

"I'm not sure," he then began. "All I know is…they were there…killing people. People who didn't deserve it," he said as he shook. "And now they're in the city with us…as if they're living here with us…and I can't help but think how much more dangerous it is to live in the sector," he said while Mya spoke.

"I think it's safe to say living in the city, in general, may be a hard place to live," she said as he nodded and spoke.

"I just thought things were changing…that, we didn't have to fear them as much…that if we just stay here in the sector, we would be left alone…and yet the IOBs still come and do whatever they want to us," he said as he shook. "I…I just can't believe I let this happen," he said looking out the window as Mya spoke.

"Hey…it's ok…" she then said grabbing his hands. "You're ok…you're ok, look at me," she said as he turned to her. "This isn't your fault," she then began. "You didn't do anything wrong…you did what you could. Regardless of what you would've done, there's no changing of what could've happened," she stated as Clay shook.

"But still…I was in the woods…I was distracted…distracted over myself when my focus should've been on the city," he said as she shook.

"Maybe…but it's not your fault, Clay…it's not," she stated while he slowly nodded.

Continuing to clean out his wounds, she soon then began preparing stitches as Clay could only look at her thinking while he then shook his head and spoke.

"You should've seen the looks on those people's faces when they were coming back to the city," he then said looking out the window. "They were broken, shattered, bereft of hope…and yet they still went back to the city because they knew they had nowhere else to go," he said, thinking. "I can't imagine how they feel right now. All they've ever done was try to live…live a normal life in this world…but yet when any sign of freedom comes, the IOBs just strip it away," he said as she thought. "I just don't understand…the IOBs…they just…keep coming. They keep trying to show us how dominant they are…how much better than us they are…they do whatever they can to show they are better…and for what? Because of our looks?" he said shaking his head. "Because of how we are…because of the way we live…it's like they keep trying to torture us for something we have no control over," he said as Mya shook looking at him.

"IOBs…they can be complicated, Clay…no matter what, they continue to always find a way to bring us down. And when that happens, we need to keep our heads up and stand together," she said as Clay spoke.

"I just don't know if we can stop them…stop them from hurting us anymore," he said as Mya shook.

"It takes time, Clay…but eventually, our time will come and we can fight. We can stop them from hurting anyone else…and we may not have the numbers…but we have our pride. Our strengths. Our experience that no IOB could ever understand," she said as Clay sat looking at her.

"But…what if I can't do this…what if I can't stop them from attacking the sector and its people?" he said as Mya shook.

"Clay…you are an amazing person," she then began. "What you do…what you stand for and the people you protect…You're more than what you think you are…you just haven't taken the moment to see it," she said as he thought. "If there is anyone in the world who can do this…I know it's you," she said as he sat looking at her kiss his head. Putting their heads together then, she slowly then shook as she spoke. "It's going to be ok…I promise," she said as she hugged him, while he rested his head on her and closed his eyes.

Soon as the hours began to pass, Clay found himself lying in bed while he saw Mya sleeping beside him. With night over them, he lay looking up at the wall as he thought about what happened in the city. From the alarms not being sent off, to an IOB ship appearing, he could only ask himself questions, becoming more and more curious with each minute passing by.

Eventually falling asleep, he later found himself tossing and turning in bed as within his sleep, he began reliving his memories again. Soon seeing Mya lying on the street ground, Clay quickly found himself running up to her as he kneeled down shaking her while the chaos all around the city continued.

"Mya! Are you alright?! Mya!" he began yelling as Mya slowly opened up her eyes. With her skin burned across her face and arms, she could only softly speak as Clay looked to her.

"Clay?" she began as Clay smiled.

"Don't worry. Everything is going to be ok. I'm going to get you out of here alright?!" he said to her as she nodded. Soon looking around, figuring out what to do, he looked for a path to go down on as he needed to get out of the city.

Looking back down at Mya, he then put his arm under her and lifted her off the ground. Carrying her in his arms, he soon rested her head against her as he shook his head and spoke.

"Mya…Mya look at me…" he said as she slowly looked at him. "Don't worry…don't worry about anything else…just focus on me," he said as she then nodded, while he looked ahead of him and began walking through the city.

With explosions landing all around them, and debris floating into the air, Clay stumbled around keeping his balance as he held Mya tightly in his arms. Determined to make it out, he continued to follow people away from the scene as they found themselves going down into the subway station, while he soon stopped and noticed a massive red blast in the distance. Shooting down from the sky, he watched the beam slowly descend down when it all of a sudden crashed into the city and shook the ground. Clay stumbled down to one knee when he saw buildings all around him collapse. With smoke and debris and yelling everywhere, he soon forced himself down towards the subway station with the other people, while a wave of debris shot out all across the surface…

Quickly waking up, Clay found himself sitting. Breathing heavily, he looked around to realize he was still in the cabin. Looking beside him, Mya was asleep and he must have drifted off as well.

Putting his hand over his head, he slowly then began to calm himself down while he breathed slowly. Thinking, he then looked beside him as his armor and suit were placed in the corner of the room.

Soon putting it on, he stood turning back to Mya who was still asleep. Leaning over as he kissed her on the cheek, he slowly then worked his way outside while he closed the door.

Walking out to the field, he soon felt the cool air around him while he pulled out his disk. Traveling back through Sector A, he headed down through the town streets as he hovered over vast number of people asleep.

Walking his way towards the sector's entrance, he reached over to where he had fought the IOBs as he stood with one foot down from his board. Looking at the city of Hollandview in the distance, he soon heard someone walk up behind him. He glanced back to see Emily coming over towards him. Turning away, he then spoke.

"You follow me here?" he asked as Emily shook.

"Not exactly. The guards noticed someone moving through the streets. I assumed it would be you," she said as Clay nodded while she walked closer to him. "I take it you have a lot of questions too," she said. Clay looked down thinking while she spoke.

"You could say..." he began. "It's hard to sleep knowing what happened here...knowing at any point, they could come," he said as Emily nodded.

"True...but considering the shape you were in, I didn't think you would be out here so quickly...you seemed to be in a lot of pain," she said as Clay shook.

"Yeah, well I guess you get used to it after a while," he said lightly as she chuckled.

"I don't doubt that," she then said. "I had my fair share of pain before I came here," she said as Clay thought to himself while he spoke.

"I take it you had it pretty rough before you came here," he said as she thought.

"You could say..." she then began. "I remember when all of this began...I was in a different city as I watched the IOBs from all around begin to destroy the people I loved. I was with the national guard with a couple of recruits at the time," she began as she shook her head. "I remember the countless days and hours we put into the city. We would stay underground in the subway stations, bringing in whatever survivors we could and giving them medicine, food or water. At that time, all I wanted to do was save people. But as I did, I soon began watching the people all around me slowly die...and I learned then that trying to save people made me lose more people, people I loved more than I could say," she said as she shook her head.

"After a few weeks of trying to find people, I tried to just survive myself for the next five months," she then stated. "Along the way, I watched every single person I became friends with die...as if the IOBs were trying to torture me every day...but eventually there was one day...one day when the city was heavily attacked by the IOB's...and in a matter of seconds, nothing was left standing...they destroyed everything while I was cramped up in the sewers...All I can remember was making my way out of the city, looking back at a place I had once lived in for so long...And that's the last time I ever saw it..." she said looking at Clay. "A city where I thought hope lay," she stated as Clay spoke.

"I'm sorry," he said as she shook.

"Yeah well after that...I was on my own. Meeting anyone that could help me...But when I met people...looking and begging for help, they only made things worse…and they only destroyed my life even more to the point, that I couldn't trust anybody anymore…and the only thing I could do then, was fight…fight on my own," she said as Clay nodded. "But during that whole time, I didn't know where my path was going. I didn't know where I belonged…In my eyes, people were just another enemy to me…At least, until I found Sector A…that's when I found hope again…But now all I can do is fight like I said…fight to fill the void…fight so I can put all of this behind me and find myself again. Fight to never experience the pain again," she said as Clay stood. Thinking to himself for a moment, he soon took a deep breathe in as he shook his head and spoke.

"There was a time once…back when the IOBs first invaded…" he said while she looked at him. "I was traveling underground with a large group of people to get away from the IOBs…and after a couple days we managed to make it to the surface…" he said as he looked around at the city before him. "You should have seen the look on their faces when they saw it…when they saw the city for the first time with the IOBs…they were so scared…people were panicking…" he said while he looked down. "Eventually, we began wondering what to do as people wanted to go back…they didn't believe they could survive outside the city…and others thought they were safer from within…At some point, there was a man that eventually stepped up, who wanted to head back. He tried to convince people to join him but I knew something wasn't right…like something didn't feel right…and when I tried to stop him, he immediately pushed me away saying, '"Why should we listen to someone like you? If anything, it's your people who probably did this!"' He said as I froze…" he stated. "I didn't know what to think at the time…but I know when I looked at each and every person around me, they were all thinking the same thing as they soon turned away…" he said while she stood.

"And while a few stayed back, the majority of them left…and soon after, the city was bombed by a military strike that wiped out the rest of the city in order to contain the IOBs," he said looking down, while Emily shook. "Throughout my life…I've faced a lot of criticism…criticism over who I am and the way I looked…and as a man of color, I've heard a lot…but never did I imagine, people would still refuse to listen to someone who they thought of as 'different'…even when the world was ending," he said while he was still. Thinking to himself, he soon took a deep breath, shaking his head while he looked at the ground before him and spoke, "If there was anything I learned that day…anything that I've learned throughout my whole life, it's this," he said while Emily looked to him.

"We live in a world full of diverse people…people with many beliefs…many thoughts…and at some point in time…we all learned for ourselves, how people can be just as bad as IOBs," he stated while she stood thinking to herself. "It's what we do now to change…" he said, stepping back onto his board, soon hovering off the ground.

Standing still, Emily looked back at him while she spoke.

"You want me to come with you?" she eventually asked. Clay shook his head.

"No I…I'd rather be alone. Protect the city until I get back?" he said as he soon took off from the scene and headed deeper into Hollandview.

Chapter 4

Working his way through the city, Clay hovered miles out north of the city's center. Hours passed by and he found himself looking around at the destroyed city

Eventually hovering through a city street, Clay turned to his side as he put one foot on the ground. Holding himself still, he looked ahead of him as a fallen building blocked his way.

Looking at it, he took a deep breath. He soon stepped off his board while it descended. Picking it up, he clipped it to his waist and turned back to face the building, before walking towards it.

Slowly working his way through the dark city, he went up to the fallen structure. Looking up at it, he put his hands against it.

Thinking to himself, he looked down shaking his head. "Uh, what am I doing here?" he whispered to himself while he stood there. As he did, he then heard snarling noises appearing behind him. Slowly hearing them, he turned around while he kept his eyes open.

Trying to see anything, he could hear the noises coming closer as he quickly turned back to the building before him. Noticing a crack in the structure, he quickly crawled in.

Squeezing in between the concrete, he slowly turned around, looking behind him.

Eventually as he sat, he soon watched as a pair of IOBs came around the corner and began heading towards him. Lowering himself down so he couldn't be seen, he watched the IOBs walk up from where he stood as they sniffed the air.

Searching the area, the creatures looked all around as they must have heard him passing by.

Snarling while they looked to each other, communicating in some language, Clay then watched the creatures begin to turn away.

Heading out, Clay's breathing shook while he tried to stay calm. Adjusting his hand while he tried to see where they were heading, he then accidently brushed against small pieces of debris that fell beside him.

Rolling all around, making noise, Clay quickly froze as he looked back to the IOBs, quickly looking back at him.

Seeing them notice it, one of them slowly began turning back as it looked at the building. Standing still, the creature moved its glossy eyes around while its head twitched around.

Creeping towards him, Clay held himself in the same position. Trying not to move, he stood waiting for the creature to come. Hoping it would leave, he knew it would eventually find him, if he didn't do anything.

Slowly holding out his arm beside him, he then deployed his blade while he held it close to him. Ready to strike when the creature came, Clay looked to the creature coming closer and closer to him. With the ground shaking with every step it took, it neared Clay drooling at the mouth.

With it almost near him, Clay quickly felt his body tightening up. With no other choice, he felt himself about to move and take out the creature before him when all of a sudden, he looked up to see a white bright light shine down. Glowing down at the IOBs in the street, Clay squinted putting his hand over his eyes, while the light shined.

Trying to see what was going on, Clay peeked out looking at the IOBs that were simply looking up. Seeing them standing still, the IOBs were quickly lifted up into the air, while they disappeared.

Watching them head up, Clay moved out of the concrete as he looked up to see what was in the sky.

As he did, he then watched the light disappearing over him. Looking around, he found himself looking up at a massive black figure hovering over him.

With his eyes widening while he stared up, he looked at an IOB ship towering over the city. Looking at the ship in awe, he watched as it slowly started to cloak itself with a shield. Enveloping the ship, as soon as the cloak covered it, it vanished into thin air.

Watching it head away from the scene, Clay walked out of the building debris, as he stood looking up.

Searching all around, he knew he had found something dangerous traveling around the city.

With fear overcoming him, he knew he needed to get back as he quickly turned away to get back to Sector A…

Chapter 5

Walking through the streets, Emily soon made her way back to Nick. Standing outside a massive eight story tall building in the middle of Sector A, he soon turned to her and spoke.

"Hey…good of you to make it" he began as Emily shook his hand.

"Sorry I got side-tracked on something" she replied as she looked at the building. "So the alarm station…why did you bring me here?" she asked as he spoke.

"We've got a new development on something…something that has to do with the alarm system not being set off…" he said as Emily nodded.

"Isn't it simple? I mean the soldiers were scared…it's on them for not letting it off," she said as Nick quickly shook his head.

"It's a little bit more complicated than that," he said as he began walking ahead to the building while Emily followed.

Entering, they began to look around at the wrecked structure passing empty halls and offices. Going up to the staircase, they both began to walk their way up the flights of stairs until they made it to the top floor.

Opening the door to a room slowly, Emily walked in to notice other IOB containment forces walking around the dark room as they lit up the room with lamps. Near the windows of the room, a control system could be seen connected to used car batteries. Controlling speakers were placed all across the building structure outside, the control room served as the alarm system for the city.

Looking around the room, Emily soon noticed the smell of blood in the air. Staring down at the ground, trails of blood marked the floor as she looked around to see a scene of horror.

Thinking what possible could have happened, she noticed the bodies of dead soldiers being covered up with blankets while Nick soon walked up beside her and spoke.

"Clay asked me to look into this…Now I know why. This wasn't a coincidence," he said as Emily looked down thinking, trying to keep her eyes away from the scene.

"Hum…what happened here?" she asked as Nick walked around the room.

"The soldiers we had stationed here…they were murdered. That's why the alarm wasn't set off…" he said quietly as Emily watched the soldiers start taking the bodies out of the room while they walked past her.

"Does Clay know about this?" she asked as Nick shook his head.

"No, not yet. We don't even know who did this," he said as Emily looked at him.

"Do you think…" she then began as she thought. "Do you think one of our men did this?" she asked.

"Maybe…but it's strange…if someone did this, there's no way they would have known about the IOB invasion," he said.

"So it could have happened…earlier?" she said as he nodded.

"By the looks of it…but still, I find it hard to believe anyone would do this…it just doesn't make sense" he stated as she thought.

"So what do you think happened?" she asked as Nick looked at her for a moment and soon looked at the soldiers around them. Grabbing her arm, he soon pulled her to the side of the room while he began to whisper.

"Well…I don't know if this is true…but I heard soldiers talking about someone running in to pull the alarm when they heard the IOBs invaded," he said as she looked at him. "Apparently, a female soldier walked in to find an IOB in the room…she said it killed all the men. When she saw it, she stated that the IOB ran out of the room…Now rumor has it, she then tried to warn the other soldiers…but they were too busy with the evacuation," he said as Emily shook.

"That's crazy…" she then said as she thought. "So where is she now?" she asked as Nick shook. "That's the thing…no one knew who she was. Said she covered her face up like she was in disguise or something…the only thing that made her stand out was her wearing an IOBCF uniform," he stated.

"Well, maybe we can try to track her down then…where was the last time she was seen?" she asked as Nick spoke.

"No one knows. Some people said she went to join the defenses afterwards, where she might've gotten caught in the fight," he said as she thought.

"No…that's impossible. I was there the entire time. I didn't see anyone like that with me," she said as Nick shrugged his shoulders as he spoke.

"Well, unless we find out more here…we have nothing but a rumor to go with…AND, with her gone, we really have no evidence to tell us what really happened here," he said. "Until we can come up with anything, we'll need to keep investigating…keep an eye out for anything suspicious," he said as Emily looked around at the scene and spoke.

"Well, we better hurry," she then said looking at everyone. "It's only a matter of time before people start freaking out," she said as Nick shook his head beside her.

"Oh they already have…" he simply said as he eventually walked away while Emily glanced back at him, leaving the room.

Chapter 6

From within Sector A, Liz slowly looked around while she stood outside her tent. With the clouds in the night slowly moving over her, she then noticed Clay standing in the street as Emily stood next to him. Looking at them as they spoke, he soon climbed up on his board and head deeper into Hollandview city. Seeing him head away, she kept her thoughts to herself while she turned away back into her tent.

Walking in, Holly could be seen inside, in a white t-shirt and sweatpants as she looked at Liz walking around her. With stitches on her face, Holly slowly looked at her as she then spoke.

"Is everything alright, Liz?" she then began as Liz shook.

"Uhm yeah, it's just…I saw Clay in the streets. Heading into the city," she said.

"Oh? What's he doing out there?" Holly asked.

"That's what I'm worried about," she said as she then sat next to Holly. "Any time he's forced to go in the city usually means nothing good," she said as Holly thought nodding next to her as Liz then handed her a cloth. "Anyway, I brought this for you to cover up those stitches. Want me to tape it on?" she asked as Holly nodded.

"Sure…thank you Liz," she said as Liz nodded putting the small piece of cloth on her cheek.

"Of course," she said working on it as Holly sat lost in her thoughts. Eventually taping it on, Holly soon shook her head as she looked at Liz.

"I know this may sound weird…but can I ask you something?" she asked as Liz shook. "Uhm yeah, what is it?" she asked as Holly smiled.

"It's just…I've been in Sector A for only a couple months…and since then you've never mentioned much about your past…you know, before all this happened," she said as Liz smiled as she nodded and spoke.

"I guess I've never really thought about it," she then said as she thought. "Most of the time I try not to remember because life wasn't much different before the IOBs," she said as Holly shook.

"How so?" she asked as Liz nodded.

"Back before the city was attacked, I was actually homeless. I lived out on the streets" she said as she thought. "I recently got done with college and I managed to obtain a business degree…but when no opportunities came, I ended up running out of all my money…and I ended up homeless," she said as she nodded. "It was scary at first…but after a while it was something you kind of got used to. And when the IOBs came for the first time, it really didn't change anything," she said as Holly shook.

"I'm so sorry, Liz" she said as Liz shook.

"No, it's alright…because if I didn't experience it then, I might've not been ready for this," she said as Holly shook and thought.

"I guess not all of us had it good before the invasion," she said as Liz smiled.

"It appears so…" she said as Holly shook.

"So, how did you manage to become a nurse?" she then asked as Liz thought.

"Well, it's a bit complicated there…but my mother was a nurse and she taught me everything she knew when I was a kid. She let me study her books, and let me take old examination papers she kept behind her…she was an amazing woman until one day the cancer in her eventually caught up…" she said as Holly sat. "But aside from that, when the invasion began I volunteered a lot…helping anyone I could. And it wasn't until then that I met a woman here in the city who was a doctor. She taught me some things, but I only assisted her at times…and it wasn't until she had passed away that I took her place. Ever since then, I've just been helping anyone I could and one day, I met Clay," she said as Holly nodded.

"Was he different then?" she asked.

"Not really…he was still pretty much the same," she said as she nodded. "Mostly quiet as he's a man of few words…but I remember the first day I met him…He was hurt…I think it was the first time he had ever fought against IOBs. He was lying in the streets, bleeding out practically…I waited to see if anyone would help him but people kept on walking by. I knew some people were nurses and doctors…but they kept to themselves a lot and only helped when it was really needed. So, after a while I walked up to him and I looked at his wounds. Ever since that day, we kind of just became friends," she said as she smiled.

"What can you tell me about him? I mean, people tell stories about him all the time…but yet they don't know he is…" she said as Liz shook her head.

"Well there's not much to say to be honest," she then began. "I mean he's kind and loyal…He does whatever he can to protect the city…but at the same time, he deals with things in his past. Things he can't forget…things that kind of haunt him," she said as she thought. "He's been through a lot…more than any person should anyways," she stated as Holly nodded, taking a deep breath.

"Yeah…unfortunately, I know what that feeling is like," she said as Liz looked at her.

"We all have things that haunt us in our past…but it's also what we do to move on. And I think sometimes, it can be hard for some people…" she said as Holly smiled and leaned her head on Liz. Putting her arm around her, Liz held her next to her while Holly spoke.

"I guess after all this time…I never knew how special it would feel to be here…to be with you…to be safe…knowing a life could start here," she said as Liz smiled. "You are everything to me Liz…and I know in my heart…that there's no one else I'd rather be with than you," she said as Liz shook her head and spoke.

"And there's no one I'd rather be with you than you…" she said hugging Holly while they held each other close with the night skies continuing to roll over them.

Chapter 7

Soon, the next morning came. Clay had worked his way back to his cabin as he sat at the table across from Mya. Looking out the window, he sat quietly while he shook his head and eventually spoke.

"I couldn't believe it when I saw it, Mya…it was there. I didn't know what to do…all I could think of, was to come here…I didn't want to worry the city…I didn't want them to be scared…" he said as he thought. "I knew as soon as I saw the ship, I had to tell you…" he said as Mya shook.

"No, it's ok, Clay," she then said as she held his hands. "It's going to be ok…First, we have to figure out why an IOB ship is here in the first place," she said as Clay shook.

"I…I don't know," he then said, shaking his head. "It doesn't make sense…even if the sector was found, why would so many IOBs be needed to destroy it? I…just-" he said thinking while Mya shook.

"It's ok, Clay…talk to me," she said as he soon looked to her.

"I just…I just wish I didn't know…I wish I didn't see that ship in the city…all I could think of is how scared I am and how scared I am to protect the city," he said. "And when I think about that…I can only imagine if someone in Sector A saw it…if they found that ship and the fear they would have…if they knew this was coming for them, they would all panic…they would all give up because in reality, this is something no one should ever see," he said as she looked to him. "I was taught to be strong…to be lucky enough to have found this ship before anyone…but now that I did, I…I don't know what to do," he said looking down while he thought. Taking a moment to breath, Mya soon nodded her head before him as she held his hands tighter and spoke.

"You have to tell them Clay…you have to warn the city," she said as he shook.

“But if I do…I’d be risking hundreds of lives trying to defend the city and keep order,” he said as Mya shook.

“We have to do what you need to do to protect the city, Clay…And I know you will. But the only way to do that is to tell them…tell them and help them…let them help you,” she said as he thought. “We have to fix this Clay…we can’t let the IOBs beat us…if anything, at least tell Nick,” she stated as Clay shook.

“If I tell Nick, people will want to get involved,” he said as she spoke. “Then he should,” she stated as he sat thinking while he shook.

“I…I don’t think I can do that, Mya. If I tell him…what are we going to do? We can’t attack the ship in the middle of the city…” he said.

“Maybe…but there are other options,” she said looking at him while Clay sat thinking while he soon nodded.

Moments later, Clay found himself quickly heading back down towards Sector A. Hovering down on his disk, he landed down to the streets as people in the city walked all around him. Picking up the disk, he soon began looking around as he went forward through the crowded streets.

As he began to work his way through, Emily could be seen standing ahead as she was talking to another soldier. While she did, she soon noticed Clay in the distance and went towards him to talk.

“Well,” she began as Clay looked up at her. “Back again, I see,” she said as Clay walked up to her while he looked around. “So you find anything last night?” she then asked while Clay nodded.

Walking ahead, the two went into a building to the side as Nick could be seen inside. Standing behind a desk in an abandoned office, he stood talking to a couple of soldiers before him as Emily walked in and spoke.

“Hey Nick, someone here who wants to talk to you,” she said.

“Who is it?” he said as the soldiers all turned to see Clay slowly walking in. As he did, Nick quickly nodded and dismissed the soldiers from the room. Clay stood by. Waiting for the soldiers to walk out, he soon looked at Nick as they began to speak.

"I take it this isn't good news," he said as Clay shook his head.

"Not exactly," he began. "I went into the city last night and I found something," he said as Nick listened. "What did you see?" he asked as Clay looked at him.

"An IOB ship…a warship to be exact," he said as Nick stood staring at the ground, as he leaned forward and put his hands on the desk while he thought. Shaking his head, it was as if he couldn't believe it while he looked back at Clay.

"That…that can't be possible…" he said as Clay shook his head.

"I saw it for myself…it's real. And it's out there," he said as Emily looked at them in confusion as she spoke.

"What do you mean…by an IOB warship?" she began as Clay looked at her. "Is this the same ship that attacked with the IOBs yesterday?" she asked as Clay shook his head.

"No…" he then said as he looked to her. "The ship I saw yesterday was a scout ship. It's a type of ship the IOBs send out to specific areas…targeted areas" he said.

Nick then spoke, "IOB scout ships pretty much lead IOBs to an assigned area, you could say."

Emily spoke, "Ok, so what is the warship for?" she asked as Clay stood thinking. He soon looked back at her. "Pretty much for one purpose," he said as she realized what he meant.

"To destroy things," she said as Nick nodded.

"Not destroy things…but obliterate them," he then said while she looked at him. "They terminate any kind of life that's in their way…including a whole city," he said.

"The last time an IOB warship was seen around here, was when the invasion began," Clay said. "They carry a heavy arsenal…and a weapon capable of destroying Sector A," he stated.

"Ok…" she began slowly. "So what are we supposed to do?" she asked as Nick shook his head.

"Well, they're not easy to take down. When the world had some fire power, they would aim towards the access points where the arsenal was firing from…hoping to blow the ship up from within. I've seen it done before but it's about distracting the ship in time to do it. You won't believe how hard that can be though," he said.

"But it is possible? Right?" she asked, looking at Clay as he thought.

"Maybe, but none of us has ever taken down a warship before," he began. "It would take everything we've got to destroy it," he said as she nodded.

"Well, all we need to do is try then. We should go find the ship and take it down before it comes to Sector A," she said as Clay shook his head.

"Ideally, we would. But that's why I'm here," he said as Nick nodded.

"So what are you thinking Clay?" he said as Clay thought.

"To stay here, to build the defenses and take our time with this. If we can get more people to join us, then we can start coming up with an actual plan to fight against them," he said as Nick nodded while Emily shook her head.

"You can't be serious…" she quickly began. "If we do that, aren't we putting lives at risk still?"

"Eh, I have to agree with Clay," Nick spoke up. "It's the safest option and going out into enemy territory is pretty much suicide…we can't risk it," he said as Emily spoke.

"But shouldn't this decision be based on what the people want?" she then began. "Shouldn't they decide their fate and make the obvious choice?" she asked.

Nick spoke, "It may not be the obvious choice to you, but it's our best option to protect the people. I mean if we prepare Sector A's defenses and train the right people, then we will have a greater chance fighting off the IOBs then we would have of invading them," he said. "Besides, in case something does go wrong, we'll have an evacuation ready. This time with a functioning alarm and rendezvous points set," he said looking toward Emily. "From there, then the people can decide their fate," he said as Clay nodded. Standing while he thought, Clay soon looked out into the streets as people were yelling.

Hearing them, Nick quickly walked out with Emily, while Clay followed as they stepped out looking up into the grey skies with a flare shooting off, deep within the city.

"It doesn't look too far from here," Emily said. Clay turned to Nick and asked, "One of yours?" Nick shook his head.

"Not that I know of," he then began. "We haven't had to send out any scout teams in the past week," he said as Clay listened.

"Well whoever it is, it looks like they might need some help" Clay said as Nick nodded.

"I can send out a team to check it out…" he said as Clay thought to himself for a moment before he spoke, "No, I'll go," he said as Nick looked to him.

"You sure?" he began slowly as Clay nodded. "You don't want anyone to come with you?"

"No, it's safer this way. Besides, I could see if I can find any more clues about this IOB ship when I'm there," he said as Nick nodded while Clay pulled out his disk. "Start making preparations now until I get back," he then said, stepping onto the disk as he headed over the people in the streets, and in the direction of the flare.

As he did, Emily could be seen standing beside Nick as she slowly spoke while Clay went away. "You didn't tell him about the alarm system," she said as Nick shook.

"I think he has enough on his plate right now…we all do. Besides, we got other things to worry about," he said as Emily looked at him.

"So what are we supposed to do? Just sit and wait?" she asked as Nick spoke.

"No…Start preparations like he said. We'll start gathering the soldiers for now…At least spread the word to all of them. If anything we need to be prepared…and when Clay gets back, we'll go from there," he said nodding as he soon headed back to his tent, while Emily shook her head looking back up to the faint flare in the sky.

Chapter 8

Hovering through the city, Clay quickly looked up into the sky to see the faint lines of the flare that was shot. Following its trail, he eventually found himself within the streets where he stopped.

Stepping off his board, he slowly then knelt down and saw a flare gun lying on the ground. Picking it up, he looked around while he stood up to see any signs of anyone around. Not seeing anyone, he then put the gun on his belt when all of a sudden, he noticed small drops of blood on the ground all along the street.

Looking at them, he began to follow the trail while it headed north of the city. While he looked in the direction, he heard faint gun shots rapidly echoing across the city.

Hearing the shots go off, Clay paused for a moment as he looked around. With his eyes wide open and his chest moved rapidly, he began to breathe heavily. Trying to see what was happening, the gunshots continued to echo towards him, as he stepped back onto his board while he headed towards the sounds.

Quickly moving through the streets as he curved around alleys and debris, he stopped as he turned to one side to see two soldiers yelling out in the street. Seeing them in the distance, Clay noticed they were both dressed in white uniforms and helmets, as a trail of IOBs could be seen behind them.

Trying to escape the creatures, Clay quickly deployed his blades beside him, as he hurried towards the soldiers.

Hurling towards them, the soldiers soon watched as Clay hovered past head towards the creatures.

Quickly then jumping off his board, Clay swung a blade as he fell onto an IOB. Putting a blade through it, he stood back up as an IOB screeched towards him as he pierced a blade through its head. Grunting as he kicked it back, he turned behind him to another as he cut off its bladed arm and glided a blade across its chest. Falling down, another IOB soon lunged out with its jaws open as Clay quickly lifted up his arm. Holding it up as it bit into his armor, he quickly pierced a blade through its shoulder as he then spun, cutting it down.

With the creature falling, Clay soon stood back up as he turned to the soldiers behind him.

Seeing them, he quickly walked over listening as he could hear more IOBs screeching in the distance while the soldiers spoke.

"Who the hell are you?" a bearded soldier asked as Clay shook his head, looking toward him.

"It's not important," he said as he saw a wounded soldier unconscious on the ground before them. "What happened here?" Clay asked.

"We were on a mission," the bald soldier began. "We got separated from the rest of our team and we were attacked by the IOBs…Private here got hurt and we shot a flare hoping the rest of our team would come but they haven't," he said as all of a sudden, Clay looked back as he could hear more gunshots going off and yelling. Hearing them, Clay quickly began to think as he looked back to the soldiers.

"We need to go," he said as his blades went back into his arm. "We need to get out of here now!" he said as the bearded soldiers shook.

"Where?" he asked as Clay shook lifting up the soldier over his shoulder.

"I know a place…but we have to go. C'mon!" he said, jogging away back towards the direction of Sector A as the bald soldier beside him, spoke.

"Wait! We can't leave! What about the rest of our team?!" he said as in the distance the gunshots in the distance began to decrease while the screams and pained yells echoed towards them.

"It's too late! Just go!" he said carrying the soldier as they followed.

Running ahead, they worked their way through the massive streets of the city, Clay breathed heavily. From behind, the swarm of IOBs began to pour out into the streets and they needed to get away.

But as they ran beside them, Clay looked up as an IOB soon crashed out of a window and as it shot out, pieces of glass began falling below. Crashing down from the building beside them,

Clay fell to one knee carrying the soldier as the others tumbled down covering their faces with glass falling down over them. Quickly then, the IOB beside them screeched, lunging forward as it grabbed the bearded soldier. Yelling, the soldier struggled to get out of the creature's grasp as the IOB held his head and soon pierced a blade through him.

Seeing the soldier being killed, the bald soldier stood up quickly loading his gun as he began firing towards the IOB. Firing as the bullets bounced off its skin, the creature screeched forward grabbing the soldier by the neck as it lifted him up and smashed him to the ground. Crushing his bones, the creature soon tossed the man away as Clay was still on one knee.

Screeching into the air as it soon turned around, Clay quickly then put the soldier down beside him as he deployed his blades.

Drooling from the mouth, the IOB soon lunged forward towards Clay. Running towards him, Clay threw out his blades beside him as he ran towards the creature and slid down to his knees. Sliding pass the creature as its bladed hand merely missed his neck, he slid the blade across its leg as it soon went screeching down to the ground.

Falling, Clay quickly stood back up turning towards the IOB that fell. While he did, he soon looked up to the massive building in front of him as two more IOBs could be seen smashing through the windows as they dropped down. Crashing down while the ground shook, Clay looked to see the unconscious soldier lying beside them.

Needing to get to him, Clay quickly looked at the IOBs standing up as they twitched their heads and chomping their massive jaws.

Furiously, they soon screeched towards him. Clawing at the ground, Clay calmly stepped forward as he ran to the first, while he spun slashing the blade across it.

Sliding past him, he continued running forward to the next IOB that lunged towards him. Quickly sliding to his side, the IOB soon went over him as he then turned to his stomach with his blade pointing out. Aiming it towards the creature, he soon shot out the blade while it went firing into the IOB's head.

Screeching while it fell to the ground, Clay soon stood back up as another blade loaded on his arm. Getting it ready, an IOB soon appeared behind him as it quickly bit into his shoulder.

Pulling him back as the creature sank its teeth into his armor, he quickly threw his elbow into it. Letting him go, Clay quickly turned around gliding the blade across its chest while it then fell.

With the creature falling down before him, he soon stood as he looked ahead of him to see the herd of IOBs coming towards him.

Seeing those crawling along the massive buildings of the city, while more crawled and ran towards him on the ground, Clay quickly felt his breath leave him.

With the ground shaking, Clay quickly stepped back, as he ran towards the unconscious soldier. Running up to him, he tried lifting him up as an IOB lunged into him from the side. Hitting the ground as the IOB above him chomped its jaws above his face, he quickly shoved a blade in its mouth and pulled it out.

Pushing it off him, Clay quickly then looked back to the soldier beside him.

With more IOBs coming, he slowly grunted in pain while he put soldier over his shoulder. Getting him up, he continued running away as the IOBs behind him quickly followed.

Running through the street as the swarm of IOBs covered the entire city behind him, he followed the road ahead where he found more IOBs coming his way. Seeing them come while he stood, he then looked through connecting streets and alleys beside him as IOBs could be seen there too.

With his eyes wide open, he breathed heavily as he looked all around while IOBs surrounded him. All across the rooftops and building, the IOBs managed to pin him as he couldn't escape.

Trying to think what to do as he looked around, an IOB swung its blade behind his leg while he knelt down. Dropping the soldier, he turned towards the IOBs behind him as it soon grabbed him by the throat and threw him down. Holding him down then, Clay struggled to get up while the creature drooled over him. Raising its arm up, it looked to strike it down at him as he cringed his face.

Feeling himself about to be struck, the creature then tried to throw down its arm when all of a sudden, a German Shepherd dog leapt out before it and grabbed hold of its arm.

Watching this quick turn of events, Clay looked to the IOB struggling and screeching as the dog pinned it to the ground and grabbed it by the neck.

Holding it down in its jaws, Clay then watched as the IOB furiously swung its arms around until the dog twisted its head and it went quiet.

With the creature going silent, the big massive shepherd dog dropped the IOB down slowly and turned to Clay. Looking toward him, Clay soon recognized him as he spoke.

"Lutheran," he began as the dog spoke.

"Are you alright, Clay?" he asked in low raspy voice as Clay nodded.

"I'm fine," he then said getting back to his feet. As he did, he then looked back to the soldier beside him as the IOBs screeched around them.

"He's injured…we need to get him to safety" he said.

"We'll need a way out first" Lutheran then said as Clay quickly thought, looking around. As he did, he saw a sewer hole behind him and he quickly ran towards it. Attempting to lift it up, an IOB soon dropped down beside him as he fell. Covering his face from the debris, the creature rose up screeching towards him while Lutheran quickly lunged into it and smashed it into the concrete wall of the building. Watching it fall, Lutheran soon looked back at Clay and spoke.

"I'll hold them off until then! Just hurry!" he said as Clay nodded and forced himself back up.

Running towards the opening of the hole, he soon began to try and grip the top of the lid as he tried to lift it up.

As he did, Lutheran looked to the IOBs that lunged out towards him. Seeing them come, he lowered his head while a black metallic vest then formed around his body. Forming on him, two guns deployed out by his side as they began firing at the IOBs. Shooting into them, the IOBs were quickly being pushed back while Lutheran stood looking around watching the bullets fly.

With him holding off the IOBs, Clay continued to lift up the lid. Struggling to get it off, he felt himself in pain while the veins in his arm struck out with blood running down. Eventually lifting with all his strength, he soon managed to pull off the lid as he dragged it back. Dropping it then before him, he soon breathed heavily while he looked back to Lutheran and spoke.

"Lutheran, let's go!" he said as Lutheran nodded looking back. Looking toward the soldier then, Lutheran grabbed the soldier by his collar while IOBs screeched towards him. Shooting out blasts from above, Lutheran growled trying to pull the soldier back while the blasts and creatures were crawling all around him.

Running to help him, Clay then ran out to the IOBs as he sunk a blade into its chest. Pulling it out, he raised up his arm as an IOB threw its claws at him. Shielding himself from it, he then struck the creature down and turned to another and shot a blade into it.

Tugging the soldier towards the sewer, Clay slowly walked backwards as he held out the blade beside him. With IOBs trying to get by, he quickly struck them down before they could reach Lutheran.

Eventually reaching the sewer hole like he did, Lutheran quickly dropped the soldier down into the sewer and jumped down after him.

Glancing over his shoulder, as he could then see them ready, Clay then turned around quickly.

Dropping a flash grenade before him that blinded the creatures, Clay quickly ran towards the hole and began to climb down. Grunting as he reached the bottom, he quickly then grabbed the soldier's body and put him back on his shoulder.

Crouching down as he looked around, he heard the IOBs screeching above and he quickly began walking ahead with Lutheran.

Traveling through the tunnels, Clay looked back and he could see IOBs falling and crawling down into the hole.

Seeing them fall in, Lutheran quickly then stopped as he turned around. Facing the creatures, he soon had a gun engage by his side as it released a small missile. Firing towards the sewer opening, the IOBs watched the missile head towards them as they screeched. Exploding as it hit them, the ground quickly shook while the fireball shot out of the ground.

With IOBs being blasted away as they flew into the air with the dirt and debris, Lutheran slowly stumbled around while the tunnel before him began to collapse.

Quickly turning away, he then went to catch up with Clay as the tunnel behind them slowly came down. Cutting them off from the IOBs, they continued to run ahead and didn't look back, and eventually managed to escape from the scene.

Chapter 9

After following the tunnel for some time, Clay soon lit up a match and raised it before him. Looking around the massive sewer way, he stood around looking, with the soldier still on his shoulder. He was breathing heavily. Walking beside him, Lutheran appeared and he began to speak.

"This way," he said as Clay nodded and started to follow him.

Going around for some time, Clay could smell the musky wet sewers. Darkness lay around him and no sounds but that of the leaking, dripping water echoed in the distance.

Looking around, Clay slowly could feel himself thinking as he remembered another memory.

Heading through the tunnels in the underground subway stations, Clay held Mya before him as he walked with other survivors around him.

Looking forward where he could see the light, he slowly then shook his head as he looked at Mya and spoke. "Don't worry Mya…it's going to be ok," he said as he followed the tunnels…

Opening his eyes back up, he looked to the soldier over his shoulder as he shook his head. Back in a reality he wished he wasn't in, he continued ahead as he tried to keep his mind from drifting.

After traveling for some time, he eventually found himself laying down the soldier as he lit another match. Sitting down, he grunted in pain as he took a moment to regain his breath while Lutheran sat. Thinking for a moment while he looked to his wounds, he soon took a deep breath as he slowly began to speak to Lutheran who sat beside him.

"You know, it's funny…" he began. "A long time ago…I used to work at this place…where I walked in everyday with this group of guys that made my life hell…" he began while he thought. “Every single day…every single night, I was taunted by them because they hated everything about me…wanted to just make my life seem…meaningless,” he said. “And even though I came home beaten and broken in every way, I still managed to put a smile on for Mya when I saw her…and even though I was hurt, I would shake my head and say I was alright…because in the end, I knew I had someone I could come home to…someone who actually cared for me and was worth fighting for every night…and it’s funny now…because in a way…the IOBs are like those guys in my life…they seem to always be there…always waiting for me to come,” he said, sitting down shaking his head and Lutheran sitting down nodding, Clay then looked to Lutheran as he spoke. “I’m surprised you managed to find me out here, Luth…” he said as Lutheran nodded. “How’d you know I was here?” he asked.

“I happened to be in the city,” Lutheran spoke, “something led me here…I’m not really sure what but it must have been a hunch,” he said as Clay listened. “Next thing I know, I hear gunshots and see IOBs running through the streets…and that’s when I found you,” he said.

“Well if it weren’t for you…I probably wouldn’t have been here right now,” he said.

Lutheran spoke, “Regardless…we’re here now. That’s all that should matter.” Clay nodded. “The only question is, what led you here?” he asked

“I followed a flare here,” Clay then began. “I figured someone needed help. But when I got here, I found a group of soldiers running away from the IOBs,” he said looking at the unconscious soldier. “I tried to save them…but he’s the only one I could.”

“Who is he?” Lutheran then asked.

“I’m not sure…but by the looks of his uniform, he looks like he’s not from around here,” he said as Lutheran walked over to the soldier.

“An outsider,” he said as Clay nodded.

“The soldiers that were carrying him…they told me they were on a mission,” he said as Lutheran looked to him. “By the sounds of it, it sounded important…and if I’m right, I bet there were more of them somewhere in the city” he said.

"You think it was good idea bringing him with us?" Lutheran asked as Clay shook his head thinking.

"I don't know. But I guess we'll soon find out," he said as Lutheran nodded while they sat waiting for the soldier to wake.

After a few hours had gone by, a match on the tunnel ground was put down before the soldier, and a thin line of smoke rose up in his face. Smelling it, the soldier began to move his head around and he began to wake up. They could see his brown hair and that he was very thin. With his vision slowly beginning to come back to him, he soon found himself staring down at the ground as he spoke.

"Uh where…where am I?" he began as he looked around. Seeing nothing but darkness around him, he looked at the match beside him. Giving him some light, he soon shook his head as he continued to speak. "What's…what's going on?" he asked putting his hand over his head. While he did, he watched as Lutheran appeared, walking towards him. Seeing him there, Lutheran soon sat down looking at the soldier who simply shook and spoke.

"Um ok…" he said looking at him. The soldier then looked around. "I suppose, you don't know how I got here, do you?" he asked.

Lutheran then spoke, "You were attacked by IOBs…you were unconscious at the time but you're safe now."

The soldier froze. Looking at him with confusion, the soldier slowly shook his head as he spoke, "You…you ta…you can talk?"

"Would you rather hear me bark?" he asked as the soldier shook his head nervous He had begun panicking.

"Wh…What?! What's going on here?!" he asked as above Lutheran, the soldier looked up to see Clay walking up to them. Clay stood holding his wound on his shoulder while the soldier spoke.

"Who are you?" he asked as Clay nodded.

"That's what we've been waiting to ask you," he began.

"I…I don't understand…what am I doing here?!" he asked him as Lutheran than lunged out.

"Shut up! Unless you want me to rip out your throat!" he growled to the soldier who pulled back his head while Lutheran spoke close to him.

"Wait, Lutheran…" Clay then began as Lutheran soon eased down while he looked at Clay speaking to the soldier.

"Look, there's a lot going on here you don't understand. If you want answers, you need to answer ours first because we don't have a lot of time…do you understand?" he asked and let the soldier think for a moment.

Looking back to him and Lutheran for a moment, he soon nodded his head and spoke, "Uhm…yeah, I…I guess I can do that," he said as Clay nodded.

"Alright…So, who are you?" Clay then asked.

"Uh, Jimmy…Jimmy Connell," he began as Clay nodded for a moment.

"What are you doing in the city?" he asked.

"We were just going after a ship…a ship we spotted going into the city," he said.

"You must be talking about the warship…" he began as he thought. "How do you know about that?" he asked as the soldier spoke.

"Our forces have been trailing it for some time now. Ever since we spotted it in a run-down town," he said as Clay listened. "Once we were ready, I was deployed with a small group of others to go out and find it…hoping to destroy it," he said as he shook. "But when we got to the city, we split and I got lost…I then let out a flare but then IOBs came…and the next thing I remember was falling down and hitting my head. That must have been when I got knocked out," he said while Clay nodded looking at him as he soon spoke.

"Well after you got knocked out, your crew ran into IOBs," he then said as Jimmy thought. "They were sent out to kill them…I tried to save them but you were the only one I could. If it weren't for Lutheran here, we both probably wouldn't have been here," he said.

"Wow…I can't believe they're all dead," he said slowly while he lowered his head. Thinking for a moment, Clay soon nodded as he continued to speak to Jimmy.

"I uh…I don't recognize the white uniform…what kind of force are you?" he said.

"Uhm, we call ourselves HOPE," he then began. "…it was formed by a group of soldiers, one including our Commander Barnett. He used to be a cop before this all began. But when the world did start falling apart, he began rallying people…creating an army to fight back against the IOBs," he said as Clay stood before him. "Once he had the army, he began traveling around to different towns and cities to try and see if anyone else was alive…hoping to either join them or be free from the creatures…That's how I'm here today," he said as Lutheran spoke.

"Recruiting people and fighting IOBs?" he began. "I find it hard to believe," he said as the soldier spoke. "Well we make our own weapons, our own bullets, and we train," he began. "Aside from that, we've stocked up on food and we've managed to survive just by traveling. We kind of just know what to do", he said as Clay took a moment to think to himself and he couldn't help but shake. Soon standing back up, he began walking around as the soldier, Jimmy looked at him

"So…since you're here, I take it there's people still around the city?" he asked as Clay shook his head.

"No…but there's a large number of people trapped south from here in a section called Sector A…they're being watched over by IOB Containment Forces…at least what's left of them," he said as the soldier nodded.

"It's been a while since we've heard of IOBCF. I thought they had been all but wiped out," he said to himself. "So you must be doing pretty well since you made it this long," he said as Clay looked at him.

"Not exactly…" he began as he knelt down and lit up another flare as he threw it on the ground. "We've held back IOBs before…but nothing like this…they have a warship which is the ship you were after…they're deadly and capable of destroying this city…we don't have a lot options right now. We're pretty much just waiting…preparing our defenses," he said as the soldier spoke.

"And it's only a matter of time before they come," he said as Clay nodded.

"There's a chance we could stop it…but right now, we're just doing what we can to not risk any more lives," he said as Jimmy looked at him as he thought and spoke.

"If anything…it sounds like you could use our help," he said as Clay shook his head, looking at him. "What do you mean?" he asked as Jimmy spoke.

"Well you said you have people to protect...If there's anything I can do...I could always bring the forces into the city. Help you stop the ship. I mean that's why we're here," he began. "And by doing that, we can supply weapons and food to all the people you have in the section to help defend against them," he said as Clay thought it over while he spoke.

"How many do you have in your force?" he then asked Jimmy.

"Uh, about 800," he stated. "They're stationed east of here...outside the city," he said as Lutheran sat beside Clay.

"That's way too many to bring into the city...the IOBs will surely spot them," he said as Clay nodded. "He's right...You would have to go around into the woods to enter Sector A," he said as Jimmy nodded.

"Well it sounds like we could do it. Send out small task forces. I'm sure our Commander will agree," he said as Clay slowly stood up thinking while he looked at Lutheran. Looking toward them, Jimmy watched as they soon walked away from him to talk.

"What do you think Luth?" Clay began quietly as he looked around, "Can we trust him?"

"He seems too scared to tell a lie," he began as Clay nodded. "But even so, trusting him doesn't change the fact that something could go wrong," he said as Clay spoke.

"True, but Sector A needs the help," he said as Lutheran shook.

"Even so, what if it goes wrong? What if these people decide to destroy Sector A?" he asked as Clay thought shaking.

"Maybe but if we don't do anything, Sector A could be destroyed in a matter of days," he said as he looked at Lutheran. "There's not a lot of options here...This kid may be our only hope of defending the city," he said as Lutheran nodded while he thought to himself and spoke.

"If that's the case, we should meet with the Commander himself then. Set up a meeting to see if they really can help," he stated as Clay nodded.

"Getting there could be a problem," he said as he looked up at the tunnels. "IOBs will be tracking us," "Then we'll need a plan," he said as Clay thought to himself for a moment before speaking.

"And I have one," he said as he soon turned back to Jimmy and walked up to him.

"We really don't have any options here…so if you can help, then we really could use it," he said as Jimmy nodded. "But right now it's too dangerous to move any further through the city…we'll have to meet with them later."

"So what do we do now?" Jimmy asked as Clay shook.

"For the time being, we'll take you to Sector A, our home…and once we're ready to head out, we'll meet with your force," he said as Jimmy thought.

"Ok…I guess I can work with that," he said as Lutheran then walked up beside him. "Just know though, if this is some kind of trap…" Lutheran growled as Jimmy shook.

"No trust me! Our forces aren't anything like that. We've helped a lot of cities and towns and saved hundreds of lives…we know what we're doing," he said as Lutheran nodded and Clay took a deep breath and he held out his hand to Jimmy. Looking at it, Jimmy took it and Clay helped him up while speaking.

"The name's Clay Treston by the way," he began as Jimmy nodded to him. "I'm sorry if we got on the wrong foot…There's just a lot going on right now, like I said," he said as Jimmy nodded.

"No, none taken…If anything, I owe you my life," he began as Clay nodded and looked around as the ground shook above them while Jimmy looked around and spoke, "What's going on?" he then asked as Lutheran shook.

"IOBs…they're still looking for us," he said as Clay spoke.

"How much further until the next sewer opening?" he asked as Lutheran spoke. "Not for a couple of more miles if we want to get clear," he said as Clay nodded.

"Alright, we can't stay here forever so we need to head back now," he said, looking at Jimmy. "Can you walk?" he asked him as Jimmy looked at his leg wrapped as he spoke.

"Uhm, yeah, I think so," he said getting up as Clay then nodded. "Alright then, let's go"

Chapter 10

Screeching ahead, an IOB lunged itself towards some soldiers who shot it down. As they fell, more IOBs ran ahead, while Emily held back an IOB before her. Holding it back, she quickly stuck a blade into its head and then she pulled it out and shot towards the other IOBs coming towards her.

Eventually the IOBs fell to the ground and the soldiers ran past her while she walked up to an IOB still alive. Twirling the blade in her hand, she soon knelt down finishing it off and she stood back up slowly.

With a group of IOB bodies lying around her, she stood thinking while Nick spoke to her. "Hey," he said as she nodded.

"Ever since they came here, they've been attacking more," she said.

"And in different spots too…seems like they might be looking for weak points or something," he said.

"Whatever the reason is, I wonder if Clay is alright…" she then said. "I don't get how IOBs could have snuck around him…let alone let them in here," she said as he shook.

"Yeah well…I'm sure he's alright," he said as all of a sudden a soldier came up behind Nick and spoke. "Sir…there's a riot breaking out again," he said as Nick nodded.

"Alright, I'll deal with it in a sec," he said as the soldier nodded while he walked away. "What's that all about?" she asked as he shook.

"Ah, more problems," he began as he looked to her. "People are getting a bit antsy after all these attacks…they're wondering whether or not we can keep them safe," he said.

"Well it's no surprise considering we're here waiting for IOBs to come." she said.

"What is that supposed to mean?" he asked as she shook her head.

“Nothing…other than the fact we should step aside from Clay’s plan and actually do what is right,” she said.

“Oh no…no, no, no. We’re not having this conversation again, Emily,” he said.

“Why not Nick? Look around you…people are dying and IOBs are coming in almost every day…Are you telling me you’d rather deal with this rather than trying to end it?” she said as Nick spoke.

“Emily…you have no idea what IOBs are capable of…Now the plan is set. All we have to do is wait for Clay,” he said as she shook.

“And what happens if it’s too late…what then? What if IOBs come charging in here with their ship? What are we supposed to do?” she asked while he shook.

“You’re just going to have to trust your instincts,” he then said as he shook. “Look…I know why you would be scared…but sometimes you just have to know when to pick your battles…this just isn’t the time.” he said while she stood there thinking. “We’ll be alright…just trust me when I say everything will be okay until Clay comes,” he said, slowly walking away. She stood there thinking and looking at the IOB bodies around her.

Chapter 11

Eventually finding their way out of the tunnels, Clay found himself back onto the surface as Lutheran and Jimmy followed behind him. Looking around, they soaked in the fresh air around them while they scouted the area to see if any IOBs inhabited the area.

With it clear, Clay soon turned to them as he began to speak.

"Looks like we're in the clear. We'll need to stay low though if we want to avoid being seen. Who knows what could be lurking around here?" he said as Jimmy nodded while Lutheran spoke.

"We should rest for the moment. We have a long walk ahead of us," he said as Clay nodded while they took a moment to relax.

While Lutheran and Jimmy sat down to rest, Clay walked ahead as he went to think. Looking around the abandon city around him, he soon pulled out his ear piece as he began to speak to Mya.

"I'm sorry I took so long Mya…I didn't expect this to happen," he said to her as she replied. "No, I understand…I'm just glad you're safe…I was getting worried," she said as he nodded.

"I know…I should have told you sooner…but I needed some answers," he said while he thought. "So this soldier," she then said to him. "You really think he can help?" she asked as he spoke.

"I do…" he began. "I think with him…we may be able to protect Sector A," he said looking on. Behind him, Jimmy slowly began to speak to Lutheran while they looked at Clay.

"Who's he talking to?" Jimmy asked as Lutheran listened to Clay speaking. "His wife…Mya," he began as Jimmy shook.

"She far from here?" he asked.

"You could say that," he said as Jimmy looked at him with confusion.

"What do you mean by that?" he asked as Lutheran slowly shook his head as he looked down. "She's uh…she's dead," he said.

"Wait, she's dead?" he said looking back to Clay. "How is that possible? Who's he talking to then?"

"No one….just himself," Lutheran then stated as Jimmy could only shake his head while he spoke.

"It's kind of hard to believe," he began. "I didn't think he would be crazy," he said as Lutheran slowly shook as he looked at him and spoke.

"Clay has fought IOBs since the day this all began," he then said as Jimmy looked at him. "He fights to protect a whole city section while at the same time risking his life to save others who couldn't even tell you his name," he stated as Jimmy lowered his head. "Try living that…living in pain…living in fear while the IOBs tear away the one bit of hope you have," he said, shaking his head. "See if you do any better" he stated while they then looked at Clay walking back to them.

"Alright we should go…It's going to be a couple hours considering we're traveling by foot, so let's move" he said as Jimmy nodded and looked at Lutheran walk past him while he stood thinking.

Eventually moving on, Clay began to lead Jimmy back towards Sector A. Unable to move quickly, they walked slowly through the city with the ashes on their feet. Walking amongst the dead and the past, Jimmy could only look around at the death around him, as he soon looked at Clay and spoke.

"So Clay…how long have you been in the city?" he asked.

"About three years…ever since the invasion began," he replied.

"Must have been hard…living here," he said.

"It's complicated," he said.

"Yeah, I can imagine…" he then began. "It's amazing how long you've been able to survive though," he said as Clay nodded. "Especially with a bunch of people."

"Well…I guess we've been lucky," he then stated. "I take it the rest of the world isn't like this?" he asked as Jimmy shook his head.

“Not really…the thing that’s most different is that most of the towns and cities we find ourselves going into don’t have a lot of people living in them,” he said as Clay looked to him. “I guess people don’t stick around much,” he stated as Clay looked ahead.

“I noticed the scars across your arms, by the way” Jimmy then began. “I take it you’ve been fighting IOBs for a while now?” he said.

“More times than I can count…” he said.

“And the armor? You make it yourself?” he asked Clay.

“No, it’s from an IOB ship…I got it after I was captured,” he said as Jimmy looked at him.

“Wait…you were captured?” he asked curiously as Clay shook while he thought walking.

“It’s kind of a blur…But I do remember bits and pieces of what they were doing to me…then I mostly just remember waking up in the woods. I didn’t know what was going on as I looked around to see I was still okay. From there, I noticed the armor already around me…and that’s how I got it,” Clay stated.

“So what did they do to you?” he asked.

“…I’m not sure but after I escaped with the armor, I soon realized that my body was enhanced somehow…like I was changed,” he said. “I was a lot stronger than I used to be…stronger and faster. So much so that even my body heals on its own, faster than any ordinary person’s,” he said.

“Wow,” he began. “So you really can fight against the IOBs?” he said.

“You could say that…” he said.

“So this entire time you’ve been fighting IOBs…but you haven’t left the city?” he said.

“What do you mean?” he asked.

“Well I just figured…if I had what you had, I would do anything I could to stop the IOBs,” he said.

“Yeah well, it’s easier said than done,” he then stated and he added after a thought, “Besides…I’m still human. There’s only so much I can do,” he said.

“But if you could? Would you?” he asked.

“Would what?” he asked.

“Leave the city? Would you do it?” he asked, while they stopped. Thinking for a moment while he stood nodding, Clay soon looked back to Jimmy as he then spoke.

"Even If I could, it wouldn't make a difference…besides, I have a city to protect…that's where I need to be," he said as he soon looked away as he continued to walk.

While Jimmy stood watching him leave, Lutheran caught up to him.

"I take it he's always like this?" Jimmy asked.

"He's just scared…we all are," he asked as Jimmy looked at him.

"Even you?" he asked as Lutheran looked back at him. "Sorry…it's just that you don't strike me as the type of person that would have…feelings," he said as Lutheran shook.

"I may be an animal, but I'm still more human than you would ever know," he said as Jimmy nodded.

"So I take it you weren't always like this?" he said as Lutheran shook while they walked.

"No, I was human once, until I got captured by the IOBs…but unlike Clay, I was turned into this," he said.

"Why would they do that?" he asked.

"IOBs were new to the world, they didn't know who we were," he began. "Anyone they could get at, they experimented on. Steal our knowledge, steal our body and see what we can do to survive while we live under their torture," he stated. "They needed to show that they were superior to us," he said as Jimmy listened thoughtfully.

"I take it, it wasn't fun getting captured by them," he said as Lutheran shook his head slowly. "How did they get to you?"

"Before the IOBs came…it was like any other day," he began softly as he thought to himself.

"I remember being inside the house on the outskirts of the city…I looked up to see the lights going out above me. Looking around, I began to hear noises outside as I went to see them for myself…as soon as I opened the door, I remember looking out at the ships all overhead. People ran in the streets as explosions crashed down all around them like a warzone…I didn't know what to do. I felt myself going still as my wife stood behind me wondering what was going on. Looking out, I slowly turned to her…I looked at her as she was scared…" he said. "I wanted to comfort her so I slowly turned to her and said everything was going to be okay…but that's when it happened. When she looked at me with her eyes watering…Looked at me for protection…that's when a plane came crashing down onto our house…crushing her underneath," he said as Jimmy shook his head slowly.

"The house collapsed on me…but I managed to survive. I began crawling out of the debris out into the front lawn…my legs were broken and I had lost my arm. I was in pain trying to get out and I could only yell out her name…the name of my wife whom I loved…" he said.

"From there, I found myself looking up at the ships above me when I yelled at them. Yelling with all the strength I had left…but when I did, the plane engine on the house soon went off as it finally ignited and let out an explosion that I swear burned and consumed me," he said. "I swore I was dead at the time…but I woke up. I woke up surrounded by IOBs who managed to keep me alive…they threw me on a table…experimenting on my body as they took out bones and organs. I felt pain…pain that I was forced to live through every day," he stated. "Then one day…I found myself waking up in a cage. I woke up to find myself in the body you see here…created to be an animal. But an animal used as a weapon," he said.

"So…ho…how did you ever escape?" he asked as Lutheran shook his head.

"I was trapped for while…until one day I managed to break open the cage and tear out the necks of every IOB that was on that ship…I managed then to crawl my way out as the ship was on land…from that moment I found myself with nowhere to go. I was broken…looking for hope…looking for a reason…looking for answers," he said as he shook. "The only thing I could think of then was to head back to this city. I had a son who lived here…and I thought if I could at least find him, I could protect him from the IOBs," he said.

"Did you ever find him?" he asked.

"My memory of him was almost wiped out along with that of my wife…I couldn't remember what he looked like. I searched this whole city for weeks until I found Clay and Sector A," he said. "From there, I've been heading out into the city every day and searching to try and find him…Hoping that when I see him, I will remember him…" he said as Jimmy could only shake his head.

"I'm sorry…" he could only say as Lutheran shook.

"No…don't be," he then began. "I don't need the sympathy," he said running ahead as he went up to catch up with Clay.

While he did, Jimmy could only think to himself while he looked around the abandoned city, as he slowly continued to follow them towards Sector A.

Chapter 12

In the streets, Liz finished wrapping up a soldier's arm as he nodded and left. Watching him leave, she stood outside her tent and saw Holly sitting on the sidewalk. Seeing her, Liz slowly zipped up her jacket and walked over to her and spoke.

"Uhm Hey," she began as Holly looked up to her.

"Liz…Hi, I was going to see you after my shift," she said as Liz nodded.

"Right uhm, well, I saw you here alone and I was wondering if I could talk to you right now," she said as Holly nodded.

"Uhm, yeah sure…why? Are you alright?" she asked as Liz sat down next to her.

"I'm fine…it's just, I haven't seen you around as much. You've been on duty…" she said as Holly nodded. "I look at you these days and it seems as if something is bothering you…and I was wondering if you wanted to talk," she said as Holly thought to herself for a moment.

"I uh…I guess in a way, I've been caught up with the IOBCF a lot lately," she said looking at her. "And I thought about the other day…about what I said about starting a life here…and as much as I do…I still feel in a way like…like I don't belong here," she said.

"Oh?" Liz said.

"It's just…hard…hard to move on from losing a city…a city where I had so many friends and family. People I miss…and now I'm here. In a place I still don't know very well…around people who haven't accepted me fully yet," she said.

"It's okay, Holl…sometimes things like this take time," she said.

"I know…but it's just…at times, because of who I am, I feel like people won't accept me…they won't accept what I really am. No matter what I do, people see me as different. Especially within IOBCF…It's hard to make friends…To tell the truth to people and open up…" she said.

"Holl…there is nothing wrong with being different," she said looking into her eyes. "If anything…it's what makes us who we are today. And look at me…I'm different too…I'm still trying to open myself up to people…It's hard. But you're not alone in this," she said as Holly shook.

"But what if people don't accept you?" she said.

"Well, then those people really don't deserve to know you, do they?" she said, smiling. "And if they did have a problem with you, I'd like to see their faces the next time they need me to save their lives," she said chuckling.

"True," she said smiling at the thought. "So no matter what then, Liz…You'll always stick with me?" she asked.

"Always," she said as Holly smiled and leaned her head on Liz's shoulder while they sat in the street together.

Chapter 13

With the hours passing by, it wasn't long before Clay made it back to Sector A.

Able to see it in the distance, he along with Lutheran and Jimmy walked ahead while the guards on the building's top yelled down to Emily in the street. Looking ahead, she spotted them and she ordered the soldiers to stand down.

Lowering their weapons, Clay looked at Jimmy as he quickly spoke.

"Okay, this is it. Stay close…" he said as Jimmy nodded, while Clay looked back to see Lutheran standing from afar.

Seeing Lutheran standing back, Clay held out his hand to Jimmy to signal him to hang on, while he went back to talk. Walking up to him slowly, Clay soon stood beside Lutheran. Looking up at the snipers ahead, Lutheran soon spoke with Clay at his side.

"It's been a long time since I've been here," he began slowly. "Hard to believe how fast time can fly by," he said as Clay nodded.

"I take it you won't be staying?" Clay then asked.

"It's probably for the best…" he then began. "If anything, it's better if the people didn't see something like me around them," he said.

"You know more than I do that you're always welcome here, Luth," he said nodding. "You won't ever have to question that."

"Still…You know I was never a people's person," he said as Clay chuckled.

"Yeah…" he said looking down as he looked at him. "So where are you heading now?"

"Hard to say…I still have some personal things that I need to work on. In the meantime though, I'll try to see if I can investigate what's really going on in the city for you…give you an update if I find anything," he said as Clay nodded to him.

"Well…you know where to find me," he said as Lutheran nodded and soon stood up to turn away. Heading back into the depths of Hollandview City, Clay could only look at Lutheran walking away. He quickly yelled out to him, "Hey Luth!" Lutheran stopped and slowly looked back at him.

"Thanks again for your help…I owe you one!" he said. Lutheran looked down and seemed to smile.

"Consider us even!" Lutheran slowly said and he then turned away and vanished into the city.

Watching him leave, Jimmy walked up to stand beside Clay. "Where's he heading?" he asked.

"Somewhere he needs to be…" he said, nodding as he looked back to Sector A. "C'mon, let's go," he then said and Jimmy nodded, beginning to follow him into the Sector.

Walking back into the safety of Sector A, Clay slowly passed all the soldiers as he walked through with Jimmy. He soon came up to Emily and nodded at her, before he quickly went to see Liz.

With Jimmy sitting on the table, Liz checked over his leg and vitals to see if he was okay.

"Well, it looks like you might have a good lump on the side of your head," she began, looking at Jimmy's eyes. "But you should be alright," she said.

"Well, that's good news," he said. "And my leg?"

"You have a light cut around your calf and ankle but a few stitches should take care of that. As long as you stay off it," she said as Jimmy nodded.

"I'll try my best," he said and Liz looked at Clay.

"Hey, thanks again for this Liz," he began to her as she simply nodded.

"Yeah, of course, Clay, it's no problem," she smiled as Clay slowly pulled her to the side and spoke.

"I know this is asking a lot…but is it possible to find a place for him to stay? Somewhere safe?" he asked as she nodded.

"Yeah, I can get Holly and see if she'll set up a tent for him close by," she said as Clay nodded. "Alright, thanks Liz," he said as he walked back towards Jimmy.

"Jim, Liz here will set up a place for you to rest for the night. Until then, anything you need goes through her," he said as Jimmy slowly spoke.

“Okay…what about yourself?” he asked as Clay spoke.

“I don’t like staying in the city…I’ll meet up with you tomorrow though,” he said walking out as Jimmy nodded, while he stayed with Liz for the night.

Walking out of the tent, Clay looked around at the soldier running around him as he turned away to head back home.

As he did, Emily soon walked up beside him and they began to speak.

“So I take it things in the city went great,” she said as he slowly shook his head. “Who’s the guy that you brought?” she asked.

“He’s a soldier that I ran into in the city…he’s the one that let off the flare along with a group of others,” he began. “Unfortunately, he was the only one I could bring back…but it turns out he can actually help us,” he said.

“Help us? What do you mean?” she asked.

“He’s part of a force called HOPE stationed east of the city…they’re trained soldiers created to stop IOBs…they’re people we need who could help Sector A. So eventually, we’re going to set up a meeting with them to convince them to help us hopefully,” he said as Emily glared at him for a moment.

“You…you sure you want to do that?” she asked as he looked at her.

“Yeah, why not?” he asked.

“I don’t know…it just seems risky,” she said. “I mean, wouldn’t we be risking lives by trusting people we don’t know?”

“Regardless of who they are, we can’t turn away people who are willing to help,” he said as she shook. “Still…I just think that Sector A needs to be protected in a way,” she said.

“Maybe…but we have to consider all options and until we can speak with them…we need to start preparing ourselves,” he said walking.

“Well, we’ve already initiated the plan you wanted,” she stated. “All the soldiers have been informed of what’s happening and are being placed at their assigned positions…they’re just waiting now for your order,” she said.

“I’ll talk to Nick…see what we can come up with,” he said, looking around thinking. “Where is he anyways?” he asked.

“Busy, I guess,” she then said. “He’s working on the defenses with the soldiers last time I knew,” she said.

"Well, let me know when he's free?" he said walking away. "Why? Where you going?" she then asked.

"Somewhere I really need to be," he said pulling out his disk as he headed away from the scene.

Retreating, Clay worked his way back to his cabin to take a moment to himself. Processing everything that had happened in the last few days, he found himself sitting on a porch step while Mya sitting behind him as she wrapped her arms around him. Resting her head on his back, they sat together while the hours passed. Not saying a word, Clay could only close his eyes and enjoy the moment until night was upon them. Looking up at the sky as stars hung over them, he turned to Mya behind him as he rested his head on her lap while he slowly spoke.

"Things have turned in a different direction, Mya," he began looking around. "I feel as though there's hope again…As if seeing this soldier makes me believe in something," he stated as she looked at him

"Like what?" she asked.

"Like help is really out there for us," he said thinking to himself. "As if maybe we're not alone this whole time," he said.

"It's just hard…hard to believe I guess," he slowly then began. "Help has never come…and when it did, things only got worse…I can't help but think it could happen again," he said.

"Maybe…but you have to believe that maybe this time it's different…maybe this is something that can really help us," she said he looked up at the clouds above him.

"Yeah…maybe…but I guess we'll never know the answer unless we try," he stated. "Sometimes we don't always have the answers to things, Clay…sometimes we have to figure them out for ourselves," she stated. "Everything happens for a reason…" she simply held him tightly. "But everything is going to be ok…I promise," she said as he nodded, her hands clutching his head while he lay in the night.

Getting some sleep, it wasn't going to be long before Clay would be awakened…

Sleeping under the late night sky, he felt a breeze pass through him as he slowly woke up. Feeling himself waking up, he looked around to find himself still outside as Mya stood beside him. Rubbing his eyes as he took a deep breathe, he looked at the ground.

Feeling himself freeze for a moment, he felt the ground shake just as he looked around. Feeling the shaking, Mya stood beside him and then he walked off the porch.

"Clay…you alright?" she asked as Clay shook his head, thinking and trying to wake up fully. Shaking off his sleep, he soon looked towards the woods before him as a trail of smoke could be seen rising up in the night.

Noticing it, he shook his head as he ran off towards the woods while he heard Mya call out to him.

"Clay! Come back!" she cried as he simply continued into the woods to see what was going on.

Quickly running, he went traveling through the woods, looking ahead and breathing heavily. Trying to catch his breath, he felt his body go numb…

Standing at the edge of the cliff, he stood with his eyes wide open. With no words to describe how he was feeling, he could only stand in that place as he saw nothing but Sector A in flames.

Chapter 14

From within the city, fire began to light up the night skies. All around, the flames grew higher as the people of Sector A scrambled around the streets. Running into walls of flames, IOB containment forces quickly tried leading the people out as they headed towards the sector's evacuation points.

With the soldier's scrambling around the city, they soon looked up as they noticed something in the sky. Looking up, a massive shadow could be seen. With their eyes widening, they could only wonder what was above them. A massive screeching roar went off as they all cringed in pain and fear. Trying to cover their ears from the sound, they fell to the ground with their ears bleeding as they desperately tried to move.

Soon making it to the city, Clay descended down on top of a building rooftop as he picked up his disk. Standing back up slowly, he looked around him to see the flames in the streets. Hearing cries of the people below, he shook his head in confusion and he too couldn't understand what was happening.

But as the smoke and flames rose up to the sky, Clay found himself looking up. Breathing heavily, he soon saw something moving ahead.

Looking into the night, he saw a massive IOB-like creature in the sky. With an elongated body, it had different features than any IOB, it continued to fly over the city, with the people cowering in fear.

Seeing the creature, Clay could only stare up at the sky, shaking while the creature moved. Trying to think of what to do, fear was overwhelming him as he tried to move. Slowly his body went numb and he realized he had to think of something to stop the creature.

Looking down below him, he could see the people of Sector A feeling the same way as he was. Still trying to escape, Clay knew he had to buy them some time.

With no other choice then, he shook his head as he quickly stepped back on his board. Looking toward the creature, he merely took a deep breath as he headed straight for it.

As he did, Emily could be seen in the distance. Lying on the roof of a building, she held her sniper rifle out in front of her as she looked at the creature. With the other soldiers on the rooftops, they began firing towards the IOB as it began screeching in the sky. Flying and gliding its way around, Emily then looked at flares shooting up in the sky.

Firing upward, soldiers from other rooftops began firing rocket launchers that shot out towards the creature. Smashing into it as it screeched and nearly fell out of the sky. The creature quickly spun its way down. Looking at the creature, troops stood still while they looked at the IOB come towards them and it lit them up in a blanket of fire.

Engulfing the building tops in flames, it soon pushed itself back up towards the sky as it continued to fly around the city.

Seeing what had happened to the first wave of soldiers, Emily quickly then aimed the sniper rifle at the flying IOB. Holding herself still, she squeezed her finger on the trigger slowly. She didn't hesitate and let out a shot.

Firing into the IOB, the bullet simply deflected off its massive head as it soon turned and looked towards her.

Spotting her on top of the building, the creature shook its head screeching while it headed for her. Seeing the creature come her way then, she quickly reloaded her gun and kept her eyes on it. Firing again, the bullet deflected off its head once more as it furiously flew forward. Getting closer to her, she quickly lifted up her head as she quickly made it to her feet.

Looking at the massive creature coming closer and closer, she pulled out a handgun and held it out before her as she began firing. Hitting the creature as it approached her, she saw herself standing still, as she watched the creature open up its massive jaws. Seeing nothing but rows of teeth, she soon closed her eyes. All of a sudden, before the creature could eat her, Clay appeared hovering over the creature.

Slowly in the night, he detached his feet from the board as he began to descend on the creature.

Quickly, he deployed the blades on his wrists as he fell down onto the creature's back with the blades piercing through.

Screeching, as he landed on it, it snapped its jaws before Emily as she fell backwards.

Watching it fly over her, she lay looking toward Clay as he held himself on to the beast.

Kneeling down with the blades digging in it, he kept his head down while the wind viciously hit him in the face. Putting his head down, the IOB creature below him quickly flew over the city, while it twirled in the sky.

Trying to shake him off, Clay held himself up as he soon looked at the creature heading towards a massive building.

Seeing it approach closer, he closed his eyes as the creature pierced through the building. Colliding with the debris as it smashed through, Clay fell back with pieces of debris hitting all over him as the creature flew right through the building.

With the building cut in half as it began to collapse. Clay quickly rolled around and he soon pierced through a window and found himself falling from the sky.

Tossing around in the sky as debris and dust fell with him, he quickly tried reaching for his ankle as he clicked a button that brought his board towards him.

Quickly coming towards him, Clay grabbed hold of it just as he nearly hit the ground. Hovering downwards, he found himself being dragged on the street as he lost his grip and he began rolling around.

Eventually stopping as he lay on the hard concrete ground, he slowly tried forcing himself up while he looked at the building collapsing before him.

Seeing it fall, he looked up above it as the IOB creature still hung in the sky. Slowly it flapped its massive wings while it let out a screech.

Forcing himself to stand up, Clay eventually made it to his feet feeling his body ache. Breathing heavily, he tried to focus while he stood deploying his blades beside him again. Soon raising them up, he held them out before him while the IOB creature snarled towards him and let out another piercing screech.

Holding himself up, Clay held himself still while the screech pierced through his ears. Lunging then, the creature then forced itself forward as it glided towards Clay.

Looking at the creature, he watched it approach him as it quickly took a breath in and let out a stream of fire down at him.

Quickly raising up his arms, he lowered his head while the fire dispersed all around him.

Grunting as the heat formed around him, he held himself up in the fire as he quickly then threw down his arms. Looking around him, he stood in the fire as he looked for the IOB.

Trying to find it around him, he soon heard a screech from behind and he looked around to see the creature's jaws open behind him. Lunging back, he tried to escape as it quickly grabbed him with its massive sharp teeth sinking into his skin and armor. Cringing and yelling in pain, the creature quickly rose back up into the sky as it held Clay in its jaws.

With the creature latched onto him, Clay realized he was over the city and the creature was taking him away. Unable to move, Clay looked around to see one of his arms free and he quickly lifted up a blade into the creature's tooth. Letting him go as it screeched in pain, Clay fell, rolling onto a building top.

Rolling across the top, he nearly fell as he held onto the ledge, yelling in pain. Barely able to hold himself up, he looked over at the IOB creature flying beside him. Looking at him, its eyes were wide open and he thought he could see something in it. Still, as he looked at it, it looked away and headed back towards Sector A.

Seeing it beginning to leave, Clay began to shake his head because he couldn't get to it. Doing everything he could to raise himself up, he knew he had to stop the creature before it destroyed the rest of Sector A.

Struggling as he grunted, he shook his head trying to think of what to do. He realized he had no choice. With the IOB creature getting further away, Clay soon took a few breaths as he looked at his arm. With blood running down it, he grunted as he tried to raise it up to the sky. Eventually, pointing it towards the creature before him, he continued to breathe heavily, while he felt his arm shaking.

Trying to keep hanging on, sweat began to run down his face as he tried to focus. Trying to keep still, he closed his eyes and tried to calm his mind.

Continuing to breathe, he began to think.

With his mind clear, he soon felt Mya beside him as she wrapped her arms around him. Opening his eyes back up, he could see her close to him. Holding him tightly, she shook her head as she spoke.

"You know what to do," she said as Clay then watched her slowly reach out for Clay's arm. Holding his arm up as she kept it steady, he soon looked back towards the IOB in the sky.

With everything slowing down around him, time seemed to freeze and everything went mute.

With consciousness of every second passing by, as the creature was getting further away before him, he slowly then pressed a button in his palm.

Pressing it, a release soon clicked back as the blade on his wrist slowly deployed outwards with sparks shooting out.

Quickly bursting out, the blade shot out across the sky as it moved across the city.

Moving quickly, the IOB creature had its eyes still on the city when the blade pierced through its wings.

It then screeched and looked back at the blade attached to its wings and at the end of it, a device was clicking. It went off and ignited the air around it on fire.

Blowing up in the sky, massive explosions shot out above the city as people all around looked up to the sky to see the blast.

Seeing it hit the creature from afar, Clay, still hanging on, watched the explosion. With no movement after the hit, he could only see the creature covered in the explosion as it went crashing down to the ground.

Hitting the ground hard, the city shook and buildings collapsed down under and around it. With everything rocking, Clay felt himself barely hanging on, and he quickly adjusted his grip on the ledge and soon pulled himself up slowly. Grunting, he eventually made it back to the top of the building. Breathing heavily he rested his head back, he could only close his eyes as he felt Mya kneeling down beside him.

Chapter 15

With smoke and debris lying in the streets, Clay slowly worked his way in the city where the IOB could be seen lying ahead. Limping his way forward, he breathed heavily. Sweating as he tried to walk, he soon grunted in pain and knelt down.

Taking a moment to himself, he tried to recover. Staring into the ground, he looked up to see the IOB still breathing while it slowly moved.

Seeing it still alive, he slowly brought himself back up. Getting back to his feet, he stood silently while he knew he had to finish it.

Deploying a blade out on one arm, he soon forced himself up. Barely able to walk, he began working his way to the creature.

Rocking side to side with the smoke floating around him, he soon stopped as he noticed the ground beginning to shake. Looking down at the debris of rocks beside him, they began quivering as he tried to look around in the smoke around him. The smoke began to move away and began to circle around the creature that lay on the ground. Staring at the creature, he soon watched it beginning to move. With the smoke descending, the creature's body began to shrink. With bones crunching and skin moving around, it soon disappeared for a moment within the smoke.

While it did, Clay looked on with confusion wondering what had happened, when he soon looked up to see the smoke clearing.

Staring at where he saw the creature, he soon noticed a women lying on the ground before him. Looking at her, he soon found himself still.

Lying in a HOPE uniform in the debris, Clay could only shake as he noticed Holly lying on the ground. Seeing her there, he soon felt himself speak as he worked his way to her.

"What? Holly?" he slowly began as Holly lay curled up on the ground while she looked at him.

"No! Stop!" she said with her hand out as he looked at her. "Please…don't come any closer," she said as he slowly spoke.

"But…how…how is this possible?" he then began as Holly shook as tears began to run down her face. "I'm…I'm so sorry Clay…I'm sorry…please…please don't hurt me," she said while Clay stood shaking his head.

"No…Holly…I—" he then said when all of a sudden, beside him a gunshot went off.

Standing, Clay quickly then felt a bullet shoot into his arm and his shoulder shook. Feeling a sharp pain in his side, he felt himself fall forward as he grabbed hold of his arm. Hitting the ground hard, he could only look towards Holly as she shook her head and yelled out at him.

"NO! CLAY!" she said as Clay lay in shock. With his arm bleeding out into the concrete, he tried to breathe.

Wondering what happened to him, he soon watched as someone stepped up between him and Holly. Looking up, he soon found himself looking toward a handgun pointing down at him and he looked at the person holding it and spoke.

"Em…Emily?" he began quietly as she stood before him. With the gun pointing to his face, she simply stood shaking her head while she began to speak.

"Hello Clay," she said as Clay slowly shook his head and spoke. "What…what are you doing?!" he began as she shook.

"What I've been waiting to do for a long time" she said staring down at him as he shook. "But…Holly!" he stated as Emily nodded, looking at her.

"I know…I'm aware," she said as Holly stared at her. "Seeing her in her true form…It's like nothing I've ever seen before," she stated as Clay spoke.

"You…you knew?" he began as she then shook.

"More than knew," she then said looking back to him. "In fact it was my idea all along," she said to him as he shook his head.

"What? What are you talking about?" he asked.

"It's simple, Clay," she began. "I haven't been completely honest with you lately…and I'm surprised you haven't caught on," she said as he only shook. "Ever since I came to this city…everything I've been fighting for has led me to this moment. A moment where I can finally have a way out of this hell…a way to finally rid of these demons that follow me," she said while he could only shake.

"Hmmm, you still seem a bit confused Clay…I think I should just show you so you can understand what really is going on here," she then said raising her arms up beside her as she stepped back. Standing before him, Clay could lay watching as her eyes began to slowly turn black. The skin on her began to turn red and black as her bones broke and expanded.

With her body growing, Clay felt himself looking up as before his eyes, Emily turned into a massive IOB. An IOB far bigger than any other IOB as her hands hung down by her side. One with claws and the other a blade. Smoke came out of her mouth as her glossy eyes looked to him as she smiled and shook.

"Perhaps this will enlighten you a bit," she said as Clay shook his head in fear while he spoke. "You're an IOB too?" he said.

"Of course I am Clay! Look around you!" she then began in a grim voice. "Everything since the day the IOBs attacked…You couldn't possibly believe this was all coincidence, did you?" she said as Clay began to think. "Everything you thought that was happening to the city…the IOBs coming, the alarm system not being set off, a warship in the city…" she said as he closed his eyes.

"It was you…" he said as she nodded.

"All this time, I led you to believe I was alright…that I was just another person like you, affected by the IOBs…you were wrong. This entire time I've been planning this all against you," she said as he shook.

"Why…why are you doing this?!" he asked.

"Because," she then said looking down to him. "Without you…I can move on with my plan. A plan I've developed since the day I entered this city…"

"You see…before all this began. I was captured after my city was attacked. I was brought in by the IOBs. They brought me in and did things to me…things you are probably familiar with," she said. "But they did these experiments for a reason…they were trying to build a new species. A species that was a mix of IOBs and humans. They wanted to make a hybrid to ensure the legacy of their kind…So they tested on me and other humans. Most of them died except for me and Holly. We were the only ones able to survive their torture and that's how we met. Afterwards, once the process was complete, we were ordered to go on a mission…we were sent down to cities and towns where me and Holly were forced to kill everyone there…exterminate life…but something went wrong one day. One day when we were sent to a city, the collars that we wore, that allowed them to control us broke and we saw the ship crashing down. At that moment we knew we were free and so we ran," she said as she began walking around. "But though we were free, we knew it wouldn't last. We went into hiding as an IOB warship was sent off to find us. For months, we evaded the ship. Moving from city to city. Town to town. Going where ever we needed to go until the IOBs came…For a while we thought we were safe, but when the IOBs came they only destroyed our hope…It wasn't until we met a group of people outside this city that told us about Hollandview. They agreed to help us get here…but we were betrayed when they tried to use us…" she stated while she thought looking toward Holly. "They took something that day we thought we would never lose…they stole our lives and the last bit of hope we could possibly have…So we killed them. We killed them and we continued by ourselves to the city. And every time, we found someone trying to help, they would all try the same thing on us until we crushed their skulls into the ground" she said as Clay shook. "Eventually we did make it to Hollandview city and we were able to find Sector A…so much hope…so much freedom…it was unlike anything we'd ever experienced before," she said. "But as the days passed, Holly eventually found love. While I felt alone. I didn't know how to feel…it all didn't seem real. No matter how hard I tried to relax, I couldn't help but think about the fact the IOBs were looking for us…that they would eventually come and destroy the sector if they knew we were here…and that's when I devised a plan. A plan to longer fall at the knees of neither humans nor IOBs any longer," she said as Clay grunted in pain as he held onto his arm.

"I knew that with Holly by my side…we could do something great…at least until I found out about you," she then said looking back to him. "Slowly, I learned about you…how you were this savior…a beacon of hope for this city…you stalked my every thought and no matter what I thought, I knew deep down inside you would become a problem," she said as she slowly kneeled down before him. Looking at her, Clay could only look at the monster's face before him as she snarled her teeth and breathed over his face.

"And that's when I acted…" she began. "The alarm system, IOBs coming into the city, an IOB battleship here, and Holly turning on the city….this was all my plan," she said as Clay looked down at the ground, thinking while she spoke. "A plan you fell for," she stated as he slowly breathed in pain. "I created the diversion with the alarm system. I had Holly go in disguised and take out the alarm ringers. Then without warning, I brought the IOBs here…I let them track me here so you could face them. So they would kill you…" she said as Clay set his head on the concrete while he spoke.

"But it didn't go as planned…" he said as she shook.

"No, it didn't," she then began. "Things took a turn…you survived the attack and then you ran off into the city and found help…help that I couldn't allow. So that's when I sent Holly out to kill you…and even then you managed to avoid death somehow," she said as Clay shook his head. "It's a shame things had to end this way, Clay," she then began. "For a second, Nick made me really believe in you…believe that you could survive anything…at least until I took care of him myself," she then said as Clay froze while he looked at her.

"No…what did you do?!" he asked.

"What I had to do…" she slowly said to him. "He was weak…but he began to catch on," she said as Clay lowered his head. "You see, Nick investigated the alarm system…he later found out his troops reported that a soldier had snuck in and killed them…But he didn't tell you because he wasn't sure…he wanted to know. And when you left the city to check out that flare, he became scared…he didn't know what to do…he didn't know what was going on…but it was only a matter of time before he did…and when he did, I knew he would crawl to you…so then I killed him…knowing he could disrupt my plan," she said as Clay shook his head in frustration.

"No…no he had nothing to do with this!" he said.

"We all have something to do with it!" she began. "And in the end, we all have our role," she said as Clay looked at her.

"Why?" he began slowly as he still grunted in pain. "What could you possibly earn from this?!" he asked as Emily shook.

"The only thing I could possibly earn…respect." she said, looking down at him. "Respect and a chance to change," she said as Clay closed his eyes.

"Is this what this is all about?! Respect?!" he began as she nodded.

"Of course it is, Clay," she then said taking in a deep breathe. "That's what all of this is about. You see I lost everything because of the IOBs. No longer should they be able to breathe and live on this planet. But in the end, people proved to be just as bad, just like you said…and they too deserve to pay."

"Then what? What could you possibly do afterwards?" he asked desperately.

"I'm going to wipe out the IOBs once and for all…use the people in this city and the world to fight with me," she began. "And from there, I'll take over…I'll control them and make them do whatever I want," she said as Clay shook. "In the end, I will be their savior…I will be everything they needed…and when the world turns back to its normal self, they'll all learn from me…they'll change…change to be like me when they know my pain and I can finally live in a world where I belong," she said as Clay spoke.

"No, you'll kill them…you'll kill them all," he said as she shook.

"Then when I do, I'll find more…and when they die, I'll keep on finding more until every last person standing fights for me!" she cried as Clay looked to her. "You see Clay…in the end, it doesn't matter…because no longer will I fear waiting for the IOBs to come…no longer will I have to be looked down upon as an experiment…no longer will I have to be looked down upon by any man ever again," she stated as Clay shook. "From now on…there is no world that is free…soon every last person will know my name and fear what it means to really have pain," she slowly said to herself as Clay grunted before her while he soon spoke.

"If that's the case…then you're no different than the IOBs themselves," he said looking into her eyes as she soon nodded her head as she spoke.

“If being an IOB is what it means to gain respect…” she then said as she quickly put her clawed hands around his head and lifted him up. “Then so be it,” she said as she quickly then pierced her bladed hand into his stomach.

Grunting as the blade went into him, Holly yelled out at Emily before her as Clay gasped for air. Soon pulling out the blade from him, she slowly then lowered him to the ground and spoke in his ear.

“Your death is meaningless, Clay…” she began slowly as she looked at him. “And soon you’ll fade away from here, like the city itself,” she said as Clay breathed heavily.

Lying on his back as he tried to move, she then held the blade to his neck with her eyes looking down at him.

Threatening to finish him, she watched as all of a sudden, a series of devices rolled out towards her.

Quickly, they went off with a series of flashes before her.

Blinding her vision, she screeched and stepped back, putting her arm over her eyes as Lutheran appeared beside Clay.

Looking at him, Clay simply said his name as Lutheran spoke beside him.

“Quickly Clay! Grab on!” he said as Clay nodded and with the last of his strength left, he put his arm over Lutheran’s neck. Helping him up, Clay lay nearly unconscious on Lutheran who stood lifting him up on his back. Standing, Emily quickly recovered as she moved her arm away and looked at him.

Noticing her, Lutheran stood still with a barrel appearing beside him and it shook as it shot off a grenade towards her. Blowing up as it hit the ground, the barrel lowered as Lutheran looked away and noticed Holly lying in the debris. Staring at him, her eyes filled with tears as Lutheran simply turned away and quickly ran away from the scene.

Escaping with Clay on his back, Emily stood as the smoke formed all around her. Changing back to her normal state, she simply watched as Lutheran headed away from the scene.

“With you gone, Clay…nothing will now stand in my way,” she said smiling as she stood standing still, with the city of Hollandview at her back…

Chapter 16

Standing hours later, Clay stood within the woods and he could see the city in the distance. At the ledge of the cliff, he stood silently with the wind blowing across his hair. With stains of blood and ash on him, Liz stood beside him as she nodded and turned away. Walking down the slope he stood on, she eventually worked her way to the bottom where Lutheran and Jimmy stood. Making it to them, she soon began to clean off the blood from her hands with a bottle of water as she spoke.

"Um well…he should be stable," she began lightly as they looked at her. "I managed to close up the wound on his stomach and extract the bullet from his arm but…" she stated as Jimmy looked to her.

"But what?" he began as she shook looking at Clay.

"He's uh…he's hurting…" she said slowly as Lutheran nodded, walking up to her. "We'll take him from here…thanks Liz," he said as she nodded.

"Of course," she said as she looked down. "Just make sure he's ok, alright?" she stated as Lutheran nodded.

While she did, Lutheran looked at her leaving when he slowly called out at her.

"Liz," he began as she then stopped and looked back at him. "Be careful while you're in the city…especially around Holly. She may not be who you think she is," he said as she thought to herself about what he was saying.

"No, she's exactly who I thought she was…strong," she said to him as she turned away and began walking back towards the city.

While she did, Lutheran lowered his head back towards Jimmy.

"So…" Jimmy began looking toward Clay. "You really think he's going to be ok?" he asked as Lutheran.

“I’ll talk to him….just wait here,” he said as Jimmy nodded while Lutheran turned towards Clay and began to walk up the slope.

Slowly, Clay knelt down as he looked in the distance at the city. Lost, he watched as he saw Mya come and stand beside him as she too kneeled down, with her arms wrapped around him and spoke.

“It’s gone, Mya…everything we’ve worked so hard for,” he said as she looked to him. “All the pain we’ve had to go through…all the choices we’ve had to make…it all doesn’t matter now,” he said.

“The city may be lost, Clay…but you’re still here,” she said as he shook. “You’re alive…and this can only make you stronger from here,” she said.

“But where? Where do I go from here?” he asked.

“That’s up to you to decide, Clay,” she began as she brushed her hand over his head. “What do you do to make things right?” she said as he thought to himself. While he did, Mya disappeared. He looked to find her gone. Lutheran walked over beside him. Sitting down as he looked at the city, Clay took a deep breath as he looked on.

Clay continued to look ahead at the city as Lutheran sat down beside him and nodded his head. Looking at the city as the wind brushed across Lutheran’s fur, he sat thinking to himself and he looked at Clay who spoke.

“I remember when I first came up here,” he began slowly. “I stood here with Mya…It was the first time I had seen the city from a distance…it was beautiful…regardless of what was happening…I remember coming up here every day to clear my mind…” he said as Lutheran looked on, thinking. “My whole life…has always been a living hell…everyday, I’ve spent trying to survive while I was beaten for who I was and what I believe in…and even as a grown man…I still find myself living through that hell,” he said as Lutheran stood listening to him. “This is all my fault…I did this…none of this would have happened, if it weren’t for me…and in the end, I’ll never be able to take it back,” he said looking down. “None of this should have happened…” he said as Lutheran spoke.

“Maybe so…but it was only a matter of time, Clay, before this happened. Even you cannot deny that,” he said.

"But to fall like this…to fall under Emily…" he said shaking his head while Lutheran looked at him. "The city is lost…and I wish I knew…" he said, thinking as he shook. "I just wish I knew what to do right now…because I really don't know," he said as Lutheran looked at him, as Clay lowered his head and closed his eyes as his mind ran wild, thinking about what to do. As he did, Lutheran soon took in a breath as he looked up into the sky and spoke.

"There was a time once…a time when I trusted someone. He was a survivor…roaming, running to anywhere he could go to get away from the IOBs…I joined him after I escaped. We were traveling together and we became good friends. But a day came…we went into a small town. Survivors were stationed there. Had supplies, food, shelter…everything you could possibly think of…But one day, the person that I trusted betrayed me. He chained me to a building…and he locked me there while he murdered all those people and took those supplies for himself…and went on his way," he said as Clay thought. "I watched all those people die…heard their screams around me…it was miserable. I was forced to starve and rot in my own filth at least until you found me," he said as Clay nodded slowly while Lutheran shook. "But I blamed myself for that moment…I blamed myself for not seeing it coming…but I learned over time, that no matter what I did, there was nothing I could have done differently…and you have to accept that too," he said while Clay shook looking toward the city as he slowly spoke.

"Regardless of what I do though," he then began lightly. "I'll never be able to forgive myself for this," he said, walking away while Lutheran closed his eyes. Thinking, Lutheran soon stood up as well turning away to follow.

Walking away, Clay began limping down the slope as Jimmy looked at him. Staring at him, he tried to speak to him while Clay walked past him. Seeing him leave, Jimmy spoke.

"Hey, wait Clay…where are you going?" he asked as Clay stopped. Thinking for a moment, he soon glanced over his shoulder and he spoke to Jimmy.

"It's over…the city is taken…There's no reason to be here anymore," he said as Jimmy shook in confusion and Lutheran stood.

“I…I don’t understand,” he began as Clay turned to him. “Don’t understand what?” he began as Jimmy shook.

“About this whole thing,” he said as he shook his head. “I don’t understand what’s going on with you right now…there’s people still trapped in the city down there…people that will be driven to their death…you’re just going to give up now and let them walk?” he asked as Clay shook while he thought.

“There is no convincing me to stay…” he said.

“But this isn’t something to be convinced about, Clay,” he stated. “Those people need you!” he said as Clay looked at him lifelessly. With his eyes directed towards Jimmy, he could only shake his head while he spoke.

“I’ve done enough…there’s nothing that we can do now,” he said as Jimmy shook while he stood.

“I can’t believe this…you can’t do this…we can’t leave them, it’s not right,” he then cried out. “I may not know much about you Clay…let alone fighting…But I know that you can’t just give up on people who still need your help…regardless if you fail or not…” he said as Clay shook looking at him.

“You have no idea what’s really happening here,” he then began. “No matter what I do…it will only lead to more suffering…” he said as he looked at Jimmy. “It’s over…Lutheran will take you back to your camp and from there, you should get as far away from here as you can,” he said as Jimmy thought while Lutheran slowly walked up beside him as he spoke.

“Let’s go, Jimmy…this isn’t the place to be right now,” he said as Jimmy could only feel himself shake his head while he looked directly at Clay. Looking at him, he began to speak while Clay stood before him.

“You know when we met…you asked for my help…” he then stated. “You talked about protecting people. Protecting what was left of your city and I gave you my word that I would do whatever I could to help you…do whatever I could to get my force to help you,” he said to him. “I’m not going to give up on you, Clay…and I won’t give up on people who I know need our help…” he said as Clay stood silently while Jimmy shook with emotion. “Don’t give up on us yet either…please,” he said and turned away. Lutheran stood looking at Clay who shook his head as he looked around. With no more words to say, Lutheran then went back to follow Jimmy while Clay could only stand there thinking to himself, about the city he had lost…

Part 2

Chapter 17

One Month Later
Slowly the shower of light rain began to fall down onto the city. Where it touched the ground, the once-known section of Sector A barely stood. With buildings shattered and burnt, the people of Sector A walked around the streets and tents that were set up all around. Shivering in the cold, they tried to stay warm while the weather was beginning to change. Wrapped in blankets and worn clothes, they couldn't help but stay in their tents, hoping and wondering for help to come.

While the people of Sector A continued to try and survive, Liz could be seen standing in her tent. With a pair of jeans on and a black jacket wrapped around her, she looked at the patients lying in beds that were covered in layers of blankets. Coughing and struggling to move, Liz slowly lifted an old man's head as she helped him drink from a bowl of water.

As she set his head back down, she looked behind her and noticed people walking all around outside while they spoke. Seeing them pass by, she soon stood still as Holly came slowly walking in.

"Holly…what are you doing here?" she asked while Holly stood before her in her uniform.

"We need to go Liz…It's important," she said while Liz shook while she thought. Then she spoke, "Uhm Well, Holl…" she said looking at her patients as she walked up to Holly. "This isn't a good time…" she said as Holly spoke.

"It's Emily, Liz…" she then began looking at her. "She has news," she said as Liz looked at Holly while she nodded.

Eventually walking out, the two began working their way up the street. Looking ahead, Liz soon noticed the people of Sector A gathering together. Standing as they looked ahead, Liz slowly worked her way up to them. As she stood in the back, she watched Holly split away from her as she went to join the other containment soldiers to the side.

Liz could only stand while she looked ahead to see Emily before the people. Climbing up on a pile of debris, she slowly rose up looking down to the people of the city. Nodding her head, she had a smile on her face, she soon looked beside her.

Walking up beside her, containment force soldiers carried up a body of an IOB as they then rolled it down beside Emily. Sliding down, the people of Sector A began to yell and scream while they backed away from the dead creature's body in front of them.

Seeing the creature on the ground, Liz looked back at Emily who began to speak.

"People of Sector A…please, hear me out," she then said as they all looked to her and went quite. "I'm standing here before you today to explain to you…that the time has come…a time to begin to rise up," she said while she looked around. "This IOB you see before you here…is only the beginning of what pain could come…they lurk outside this sector everyday…waiting for you to come…waiting for you to make a mistake!" she said. "For two whole weeks…me and the rest of Sector A's bravest soldiers have been fighting desperately to protect you and provide you with the deepest of care…but the time has come to make a decision…a decision that you must all make in order to survive," she said while Liz stood listening.

"It's only a matter of time before the end comes," she then stated. "Why wait now when you can have a chance to defend yourselves?" she said.

"So what are you suggesting?" A man in the crowd asked while Emily looked at him.

"The only thing that I can suggest," she then began. "A way out…a way to help you, the people of Sector A…fight back and find a new home…away from any IOB," she said as people began mumbling around her.

"I know…I know. It sounds complicated," she slowly stated. "But I tell you the truth when I say that we can rise up together and beat the IOBs," she said.

"How can you talk about fighting when all of us are dying out here?" A woman quickly spoke out as Emily nodded. "For two weeks we've been starving…we have nowhere to go now that Sector A is gone…all the food and supplies that we had got burned up in the fires…if anything, you should be helping us find a way out of this city!" she cried out as people began to agree with her. After hearing them talk, Emily soon raised up her hand while she spoke.

"And that's what I am offering you," she said looking at her. "What I'm offering all of you here…you see, no one is going to get you to safety on their own. No one is going to fight for you so you can wait in the streets while hundreds of us fight dying to simply protect you," she said. "Don't you see? If you want a new world…a new life…you have to be willing to earn that…earn that right and fight for a life you all desire," she said while another man soon spoke up as Liz looked at him.

"So what do you suggest?" he slowly began. "Not all of us are fighters…we can barely protect ourselves" he said as Emily chuckled.

"Well you won't have to worry about that anymore," she then said as a soldier beside her handed her small glass tube. Holding it up in front of the people, Liz noticed a container of black blood inside it as Emily continued to speak. "What I'm holding here in my hand…is something far greater than anything you could all imagine…something that can change your life in an instant and make you simply…invincible," she said as a soldier stood beside her.

Taking out her knife, she then held it up to make sure the people could see it before them while she spoke. "This tube here contains blood of an IOB," she said as she looked at the soldier beside her again. "And with one drop of this blood, you can become something incredible," she said and then she stuck the knife into the soldier's stomach where he stood. The people started yelling out and she slowly then took out the knife while the stabbed soldier stood still standing.

Waiting for something to happen, the people in the street watched as the soldier's knife wound slowly began to heal. Watching his skin heal, Emily simply smiled while she looked at the people's disbelief.

"I don't lie, Sector A, when I speak," she then began. "You may not be able to fight…but you definitely can live…this blood will cure you and make you stronger. You will survive with this blood in you and I guarantee you, if you take this…then you'll be able to survive," she said while Liz looked at people thinking together quietly as more soldiers began carrying out crates. Filled with guns and knives, they set the boxes down in front of the people as Emily spoke.

"And with that strength inside you…you will be well provided for. I may not be able to promise you food…but I can promise you weapons," she said as the people looked into the crates. "With bullets and knives forged from the bones of IOBs…they'll pierce through their armor like nothing before…and in that moment, you will realize just how powerful you are compared to those creatures," she then stated as Liz shook her head. "And not only will they help you face those creatures…but no longer will you have to live in a world of pain again…I promise, if you fight for me, then I will lead you to paradise…I will lead you to where you want to be and together we can rebuild this world…a world where we can be safe again!" she said as people began to be persuaded.

As they did, Liz could only think to herself while she looked at the desperate people before her. Glancing beside her, she looked down the alley to see Lutheran standing there. He then turned away. Noticing him there, Liz quickly looked ahead towards Emily distracted before the people as she slowly split from the crowd.

Holly noticed Liz leaving and she could only think to herself while she looked back at Emily.

"It's up to you, people of Sector A…die fighting for you future…or die staying here…make your choice," she said looking at them as she then looked to Holly and simply nodded to her while she looked at the people at her feet.

While Emily continued to speak to the crowd outside, Liz stood with Lutheran. Nodding, Lutheran soon began to speak while he looked around.

"What's happening here Liz?" he began as Liz spoke.

"It's Emily, Lutheran…she's completely taken over the city section. She's trying to gather them together to help her fight against the IOBs," she said as Lutheran shook.

"What do the people think?" he then asked.

"It seems they're beginning to side with her…it's not surprising considering they would rather die fighting than starve to death" she stated as Lutheran nodded.

"And the IOBCF?" he then asked.

"The same…" she began. "It's strange. Somehow Emily convinced them to follow her. It's as if she's controlling them," she said as Lutheran thought as she spoke. "And to make things worse, they've been acting differently," she said as Lutheran looked at her.

"What do you mean?" he asked as she shook.

"I don't know, to be honest…I just know that one day I worked on Holly after her fight with Clay, her wounds healed in a matter of hours…" she then began. "It makes sense considering what Clay and Holly have in them…but what doesn't make sense is when I worked on injured IOBCF soldiers, their bodies healed the same…" she said as Lutheran thought to himself for a moment.

"Her blood…" he then began as he looked at her. "Emily must be giving her blood to soldiers…giving them enhancements to make them stronger," he said as Liz shook.

"Yeah, I heard her talking about it with the crowd…but what does she mean?" she asked as Lutheran looked at her.

"She's building her army by making them like Clay, but worse," he began as he shook his head. "…their bodies have enough of the IOB DNA in them to at least convince them to join her…but what they don't know, is the IOB blood in them will change them…make them far more dangerous than what they really are," he said as she looked at him shaking her head as she spoke.

"And that's why they must be joining her…that's what she means by making them stronger…" she said as Lutheran nodded as she spoke. "So what do we do?"

"We don't have much to work with…we'll have to develop a plan in order to stop this. If they encounter the IOBs, they'll still be wiped out in a matter of moments," he said as Liz shook.

"Emily doesn't seem to believe that," she said.

"Well when people have a purpose to fight…they can be stronger than ever," he began, looking down, thinking. "I'll have to head back for now. At least until we know what to do…In the meantime, you'll have to join her and follow her plan. It's the safest option," he said as Liz spoke.

"What if she catches on to me?" Liz then asked as Lutheran shook as he stood up on all fours. "You're a strong woman, Liz,"

Lutheran then began. "If anyone can take care of themselves it would be you…but the true threat may not be against Emily…" he said as she nodded. "Holly…you'll need to be careful around her…" he said as she nodded.

"I know…I'm still working with her…she's shut me out a bit, ever since I found out about her and Emily" she stated as she thought. "…it's going to take some time if I can gain her trust…But I think I may be able to convince her to join us," she said as Lutheran shook.

"If she does…we may still have a chance to stop Emily," he said as he then turned to his side and showed her a radio attached to his vest. "In case something happens…whether it's Holly, Emily, or yourself…just let me know and I'll try to come," he said as she nodded and quickly knelt down to give him a hug. Chuckling, Lutheran then smiled as Liz stood back up as she looked to him.

"Thank you, Lutheran," she began as he soon nodded and began to turn away. While he did, Liz spoke.

"Lutheran, wait!" she began as he slowly turned to her. "Clay…tell me, is he ok?" she asked as Lutheran looked down, thinking for a moment while Liz stood. Simply looking back at her, he shook his head slowly while she then watched him begin to turn away to escape from the city.

As Lutheran left, Liz stood standing putting the radio he gave her into her coat pocket. Taking a deep breathe in then, she soon turned around as Holly could be seen walking down the alley as she spoke.

"Liz," she asked walking up to her while Liz looked around. "Hey Holly…what's wrong?" she asked as Holly spoke.

"That's what I was going to ask you…are you alright?" she asked as Liz nodded. "Uhm yeah…I'm fine…I just needed some space," she said as Holly slowly nodded. "I heard talking…were you with someone?" she asked as Liz shook.

"No…I'm fine. I guess I was just talking to myself," she said as Holly nodded.

"Oh ok…" she said as all of a sudden, behind her, Liz looked to see Emily walking towards them. "Well this is a pleasant surprise," she began as Liz and Holly looked up at her. "Liz, it's good to see…how have you been?" she asked.

“I’ve been good…yourself?” she asked as Emily looked at her and spoke.

“I’ve never been better,” she said as Liz nodded looking away as Emily began to walk around and speak.

“So Liz, did you not like the speech?” she asked as Liz shook.

“No…I did. It was…encouraging, you could say,” she said as Emily stood.

“Well, I’m glad you liked it…” she said while Liz nodded. “By the way, I was hoping that you would help me with the injections I have planned. No one seems as qualified as yourself to do it and I’m sure more and more people will join with each passing day. So what do you say, have any free time?” she asked while Liz thought to herself.

“Of course. I can do it,” she then said as she looked at Holly. “If you’ll excuse me, I should get back to work…I have people that need me,” she said walking past Emily who stood by. Holly could only watch Liz walk away slowly as Emily soon smiled and looked at her.

“She’s a strong woman,” she began as Holly nodded and looked at her. “She seems to be taking our secret okay,” she stated.

“You have no idea what must be going through her head right now,” she said as Emily nodded. “After she found out…she hasn’t looked at me the same…she’s acting scared around me,” she said as Emily walked up to her and put her hand on her face.

“She just doesn’t understand you,” she said putting her head on Holly’s as Holly stepped away. “No…you don’t know her. Liz is everything to me. Ever since we came here to Sector A, I’ve never been happier…but then you came up with your plan…you slowly destroyed the things around you including me…You run scared…and can’t accept the fact that maybe we don’t have to fight anymore. That we can enjoy life for a moment,” she said.

“Life doesn’t end with you being happy,” Emily said as she looked at her. “In the end, pain and suffering continue to happen in the world while you sit back and try to accept life…You forget the real reason why we came here…and I won’t let that go,” she said. “If anything, Liz has compromised you…she’s made you pathetic…and she’s holding you back…maybe it’s time I end things between you two," she said while Holly looked at her slowly.

"No..." she began nervously as Emily stood beside her. "No we had a deal...you promised that if I did what you wanted me to do, you wouldn't hurt Liz," she said.

"If I remember correctly, our deal isn't over," she said walking up to Holly. "I still need you to do what you were made to do...and needless to say, do I have to remind you who really saved us from that ship?" she said as Holly stood looking down thinking. Standing, Emily soon walked up to her slowly and looked to Holly's face while she spoke.

"Don't worry, Holly...I won't do anything to your Liz," she then said as she began to walk past her. While she worked her way back through the alley, she shook her head and spoke to Holly. "Unless I have to," she said smiling, while Holly was left standing by herself. She then fell to the ground. Putting her hands over her head, she began breathing heavily as she soon began to cry.

Thinking, she soon dropped to her knees, as she felt the cold air falling onto her. Putting her hands on her head, she slowly began to cry with the tears dropping down before her as she could only feel the tear in her heart...

Chapter 18

Two Weeks Earlier

Outside, a steady shower of rain came down onto the cabin. With water dripping down from the roof, Clay could be seen inside as he lay quietly in bed. Looking out the window as he thought, he felt Mya behind him.

Wrapping her arms around him slowly, she put her head on his shoulder as she joined him watching the rain fall outside.

Thinking to himself, Mya soon spoke lightly while Clay laid.

"You seem bothered again, Clay…" she began as Clay looked down nodding. "Talk to me," she said as Clay slowly turned to her and spoke.

"I don't know what's wrong with me," he began as he looked at her. "All this time…I've done nothing but protect the city…not once did I think I couldn't do it…yet, it happened. It's gone…and I want to fight. I want to…but something is holding me back," he said as Mya spoke.

"You're scared, Clay," she said. "We all are…but we just have to do what is right. We have to fight to make things better," she said as he shook.

"Even if I could…I can't forgive myself, Mya," he began as he looked down. "I can't make up for the things that've happened…it's all my fault…and this is the only thing I can do to make it right," he said.

"Torturing yourself isn't going to help you though," she said. "Sometimes you just need to move on."

"Even if that's true…There's no changing what I did," he said thinking as he soon shook his head and got out of bed to think. Grabbing a jacket, he found himself walking outside into the rain while Mya was left behind.

Eventually, finding himself walking into the woods, he stood back up the slope as he looked at the city from afar.

Thinking to himself, he couldn't help but recall the memories he had in his head as he closed his eyes to see them.

Standing on debris, Clay remembered himself standing in the city two years ago as he looked around before him at the broken city. Standing still in his black uniform, he slowly then looked at his hand as he held a transmitter before him. He soon turned it on and began to speak.

"I…I don't know if anyone can hear me out there…but my name is Clay Treston and there are survivors here. We've been stranded in the city for over a year now. We're scared and tired. We're running low on food and water…at this point we're just wondering if we're the only ones left…we lost communication with the world a few months ago…now we're just trying to survive…The creatures…they're looking for us. They know we're here. We've taken down a few, but it's not enough…It's only a matter of time before it's too late. So if anyone is listening to this…if you do hear me…please…we need your help…we need you," he said as he slowly then put the transmitter down.

Thinking to himself as he took a deep breathe, he looked at the sun setting before him. With the warm air over him, he soon heard Nick call out to him from behind as he looked at him.

"Hey Clay…we're losing light, we should really get back," he said as Clay soon nodded and looked at the sun setting one last time while he stepped down.

Walking back to the other men, Clay handed a transmitter to a soldier as he walked ahead with Nick. "So, any luck?" he asked as Clay shook.

"No…Nothing," he began as Nick shook.

"Well, someone had to hear that" he said as Clay shook.

"Hopefully…it would make coming out here a lot easier…knowing help would come," he said as Nick nodded.

"True…" he said as Clay thought to himself. "Well anyways, we should get these supplies back to the people" he said as Clay nodded and they headed back to Sector A.

Out in the open, Clay then remembered walking back as he stood with people in the streets. Handing out food and medicine to the people, Clay stood looking around at them as Nick spoke beside him.

"Well, I think we did pretty good today…a lot better than we've done before" he began as Clay shook. "Maybe…but it won't last. We'll need a lot more if want to keep these people safe," he said.

"Eh well, we'll worry about that when we need to again. Hopefully until then, this will keep the people together," he said.

"You didn't strike me as a people guy," he said as Nick chuckled.

"Yeah, I have to admit being here for a while now has changed me," he said as Clay nodded. "I met a lot of good people here…people who need us…It's the least I can do for them," he said as Clay spoke.

"Well with you watching over them, they certainly will be safe," he smiled as Nick shook. "Ah, you really had to go there," he said as Clay shook his head.

"Nah, in all seriousness, I don't think we would have made it this far without you…you and the rest of IOBCF," he said as Nick spoke.

"Maybe…but people have someone here who really is looking out for them," he said patting Clay on the back as they stood together. "Well I think my work here is done for now…your wife, Mya, is on watch, on one of the buildings by the way. Thought you should know," he said as Clay nodded. Nick soon left as Clay looked around at the city and its people.

Soon coming back to reality, Clay stood breathing heavily as he stood and he slowly felt himself lower his head as he closed his eyes. Thinking, he soon shook his head.

"Nick…I'm sorry," he said as he opened his eyes and turned away. Walking back through the woods, Clay slowly worked his way back to the cabin and his mind continued to drift away from reality. Thinking, he soon began to remember meeting with Mya after he had spoken to Nick.

In the back of his head, he recalled himself heading to the top of a building. Reaching the top, he found Mya sitting at the ledge of the roof.

With a gun in her hand, she looked off in the distance in Sector A, as Clay slowly went up to her and spoke.

"Didn't think you were into views," he said as she turned to him.

"Hey, you're back," she said as Clay nodded as he handed her a bottle of water.

"I thought you might need a drink," he said as she looked at him and smiled and went to grab it as her hands shook as the burn marks could be seen across them, along with the skin ruined on her face as she took a sip. As she drank, Clay shook his head as he looked to her and spoke.

"Hey…you feeling alright?" he asked as she nodded.

"I…I'm ok. Just feeling under the weather," she said as Clay knelt down beside her. "You should get some rest. You don't need to be here," he said.

"No…I have to be here," she said as he looked to her. "I can't leave," she said. "Mya…" he said as she shook and looked at him.

"Clay…" she began slowly as he looked at her. "You saved my life…you carried me off while I was dying…and yet you did whatever you could to save me…and now you go out in the city everyday…you protect it every day and you fight, protecting it every day," she said as Clay shook.

"I do it to protect you," he said. "I do it for you and the city," he said as she nodded. "I know…I just…feel like I'm…like I'm useless!" she said as Clay shook. "I feel as though if I could do anything to help you…this is it," she said as Clay looked at her. "I want to protect you…but I can't," she said. "I can't do anything…and I don't want to lose you…I don't want to know I could have done something if something happens to you," she said as Clay slowly went over to her and hugged her as he slowly spoke as tears ran down her face. "Always remember something, Mya," he began as she listened in his arms. "No matter what…I will always make it back to you…" he said as Clay stood out in the woods finding himself coming back to reality.

Standing in the woods, his body shook while he slowly looked up to see Mya before him. Standing there, he could only look at her as his eyes watered, while he spoke.

"I…I…don't…what's going on?" he began to himself as Mya walked up to him. "Clay, what's wrong?" she began as Clay shook.

"I…I remember…being with you…in the city," he said as he looked at her. "We were talking on the building…you were on guard…I remember seeing you there…and you were hurting…you were hurting but you were alive," he said as she shook. "You were worried…worried about me," he said lightly as he began shaking as he stood as he continued to speak. "And after all this time…I still remember…it was the last day we hugged…the last day I saw you…as you left me…you died in your sleep…and I had to let you go…" he said with tears running down his face as she looked at him.

"Clay…what…what are you saying?" she then began as she shook her head as he spoke looking down. "I…I think you're dead," he said looking up as she was gone.

Looking around, he felt his body cringe. With his throat tightening, he kept back his tears, he could only stand looking around.

"No…No…no, no, no, don't do this…no," he began as he looked. "Mya…Mya…Mya, come back…Mya…Mya, where are you? Don't do this, please…please come back," he said, looking around.

Trying to find her, he grabbed hold of his head and shook it.

"No, Mya…no, Mya please…please…please don't do this to me…please" he said, with tears running down his face while he closed his eyes. Realizing she was gone, he eventually found himself slowly falling to his knees, as he then lay down on the ground, crying…

Chapter 19

Present Day

Miles away from Sector A, Lutheran could be seen running through the woods. Heading east, a military camp could be seen ahead. Set up along a muddy open field, tents and vehicles with the words HOPE painted in gold stood as the city could be seen in the distance.

With soldiers in white walking around, Jimmy could be seen sitting in his tent as a woman with curly black hair in uniform walked in and spoke.

"Hey, Jimmy," she said.

"Adrian…What…what are you doing here?" he asked, laughing as she smiled and hugged him. "I thought you stayed back in Burunian city?" he then asked.

"I did, but a group of soldiers were asked to come here. And so I disguised myself and went with them," she said as Jimmy shook.

"But what about your dad? I thought he didn't want you to come," he said as she spoke.

"I know…But I wanted to get out and see you. I heard you were attacked…" she said as he nodded. "Uhm yeah but…I had some help," he said as she nodded.

"So I've heard," she said smiling. "It looks like you found some new friends in the city," she stated as he nodded and thought to himself.

"Yeah…I guess you could say," he said as all of a sudden he then heard soldiers outside, yelling. Hearing them call, he quickly grabbed his helmet beside him and walked out of the tent with Adrian.

Walking out, Jimmy soon looked ahead as soldiers looked at Lutheran running into the camp. Seeing him come in, Lutheran ran past Jimmy as he went to talk to the commander.

"I better go…Nice seeing you, Adrian." he said as she soon nodded and watched him walk away.

Walking around the camp moments later, Jimmy and Lutheran walked together with HOPE Commander Lance Barnett. With a bald head and black goatee, he was a tall man in his fifties, he looked around shaking his head while he spoke.

"So…what do we know?" he began in a low voice as Jimmy shook. "Lutheran?"

"We have someone in the city who has been communicating with us…she's been keeping an eye on Emily," he began as Commander Barnett nodded.

"What does she know?" he asked as Lutheran shook.

"Apparently, troops are being gathered for war. From the looks of it, at least two thousand are looking to join her," he said as Barnett spoke.

"That's a lot of people. How is she gathering so many?" he asked.

"She's been injecting her blood into them," he said as Jimmy and Barnett looked at Lutheran as he spoke. "She's taking her blood and handing it out like some kind of drug for them to fight," he stated.

"I don't understand…what does her blood do?" Barnett asked slowly.

"Her blood contains something in it that will make people like her…once it hits the blood stream in these people, they will become stronger and faster overtime. Not only that, but their bodies will adapt to the environment around them and heal quicker until they change…" Lutheran said.

"Change into what?" Jimmy then asked.

"IOBs themselves," he stated as Jimmy shook. "Those people will be taken over by the blood…and it'll change them. Once they lose control, there is no way to predict if they'll be able to control themselves," he said as Jimmy and Commander Barnett thought.

"So they're changing into monsters pretty much," he said as Lutheran nodded.

"If Emily does convince those people in Sector A, they will fight for her…" Lutheran said.

"But would they really believe her? Would they really try and fight?" Jimmy asked as Lutheran shook.

"If you were given a way to fight for freedom, would you do it?" he said as Jimmy then thought to himself while they walked. Thinking, Commander Barnett took in a breath as he looked around.

"This situation…it's becoming a lot more complicated than we thought," he began as they walked. "First IOBs…then mutated people appearing in the city…and now hybrid people going to war…you certainly have quite the city," he said.

"You having doubts?" Lutheran asked as Barnett shook as they stopped.

"I would be lying if I said I wasn't" Barnett said.

"Well, we can't just give up on them…whether or not Emily believes she can win against the IOBs, we have to at least try and stop her…those people are being misled" Jimmy said.

"Countering Emily could be suicide though," Barnett slowly began as he thought, "If we were to attack, our forces would be annihilated by either her or the IOBs."

"Then we should counter her head on…Stop her before she drags those people against the IOBs," Jimmy said while Lutheran shook.

"And kill innocent people in the process?" he then began. "You can't expect to face her Jim without blood being spilled…they will be ready if they see us coming."

"Well, there's gotta be a way, Lutheran!" Jimmy said.

"Well, unless you two can develop a plan to stop her, I have no choice but to hold the forces here," Barnett said as Jimmy looked at him.

"Wait, sir, we can't do that. You have to give us a chance!" he said.

"Look Jim, war takes time…it takes planning. We can't just run into battle thinking you can just win," he said.

"But we're a force of eight hundred men and women," he slowly began. "If we can't stop Emily…then it will only be a matter of time before she finds us…and when they do, we'll still be going to war," he said.

"We are created to help people and stop IOBs…not end a war, soldier," he said as Jimmy shook.

"Then what are we in now?" he asked while Barnett looked around. "The world has already been at war with the IOBs…and whether we decide to stay here or not, that war is coming for us…people are going to die…and if we don't do anything about it, then what good are we? What good is HOPE?" he asked as Barnett thought to himself. "All I'm asking sir is for a chance…please," he said as Commander Barnett eventually nodded.

"I'll consider fighting…but get me a plan first. Then we'll talk," he said as Jimmy nodded while Barnett walked away back through camp.

As he did, Jimmy took a deep breath as he looked at Lutheran beside him and spoke, "Well…some progress."

"Progress doesn't win wars…if we don't think of something quick, we'll never be able to stop Emily and her army," he said as Jimmy nodded. "We'll need some more help…someone who can help us," he said as Jimmy thought while he spoke.

"Yeah but who?" he asked as he realized t Lutheran was staring at him. "You can't be talking about…" he began.

"If we want to make a plan, then we should have someone here with us that knows the city better than anyone else," Lutheran stated.

"Yeah but Lutheran…he said he didn't want any part of this," he said as Lutheran shook.

"Regardless, we don't have a choice…he's the only one we can really go to for now," he said.

"And if he doesn't? If he decides not to join us?" Jimmy asked.

"Then we'll handle this alone," he said walking away while Jimmy took a deep breath thinking. Watching Lutheran head away, Jimmy couldn't help but follow as he quickly went to grab his things to head out of camp.

Chapter 20

Within Sector A, people began lining up in the streets as Liz injected needles into their arms. With Emily's IOB blood running through them, each of their eyes began to turn black as their veins turned red.

With each person changing, Emily stood smiling as she looked at Liz under her control. While she stood, a bald soldier named Randall slowly walked up behind her and spoke.

"Commander Emily…we have news from a supply run," he began with his head twitching as his body was already changing due to the blood in him. "We intercepted a HOPE vehicle just outside the city. It was carrying weapons and supplies that we managed to occupy."

"Hmm good," she said.

"But that's not all…" he then said as she glanced back to him. "We followed the direction the vehicle was heading…and we found it," he whispered. "We found their camp," he stated as she stood thinking to herself. With the HOPE camp found, she slowly then smiled and turned to him.

"Then I think it's time they get a proper greeting," she said as he nodded.

"What are you thinking?" he asked as she shook.

"This" she then said as she unclipped a metallic black device on her waist and spoke. "I've been waiting to use this for the proper moment…so do me a favor, head to their camp. Disguise yourself as one of them and when you get there, plant this device in the ground once it's activated," she said.

"But what will it do?" he asked as she spoke.

"Well…you'll have to stick around and see," she said as the soldier soon smiled and took the device while he walked away. As he did, Emily looked back towards the streets before her. Her plans were beginning to come together.

Chapter 21

Walking out of the woods, Lutheran and Jimmy stood still and they could see Clay's cabin ahead. Looking toward it, Jimmy took a deep breath as Lutheran nodded beside him. Eventually walking ahead, they slowly made it to the cabin.

Opening the door quietly, it squeaked away while Jimmy stood in the doorway. Looking around, he walked in with Lutheran behind him as he spoke.

"Clay…you here?" he began as he looked around the cabin.

As he did, Lutheran began checking rooms while Jimmy stayed put. Nodding his head as he looked around the room, he heard the door close behind him.

Hearing it, he then turned around to see Clay.

Seeing him, Clay stood shaking his head as he spoke. "Jimmy…what are you doing here?" he asked as Jimmy shook.

"Clay…Uh, I came here to talk to you…the both of us" he then stated as Clay looked to the side to see Lutheran coming out of the room. Seeing him, Clay took a breath in, as he looked back to Jimmy and spoke.

"You shouldn't be here," he said.

"We wouldn't be here, Clay, unless we needed your help," Jimmy said.

Clay spoke, "Help?"

"The city is at war, Clay…" Lutheran stated as Clay looked down, thinking. "Emily has taken control of the people and is leading them against the IOBs…they're gathering now as we speak," he said as Jimmy spoke.

"She's been using the IOBCF for her own use and is gathering more and more people…" he began as Clay shook. "She's misleading them and giving them false hope…and they believe her".

"I already know…" he then said as they looked at him. "She told me this before it even began" he stated.

“Well, that’s not the only thing happening in the city” Lutheran then said as Clay looked at him. “Emily’s been giving her blood to the people joining her…giving them abilities like you but only uncontrollably,” he said as Clay thought. “She’s giving them a way to fight.”

“That’s why we came here…we’re trying to devise a plan and we thought you could help us,” he stated as Clay shook.

“And why me?” Clay then asked.

“Because hundreds of people are about to be killed,” Jimmy began as Clay looked away from him. “People are counting on us and unless we do something Clay…they’re going to die and if there’s anyone who can help us, it’s you” he said as Clay spoke.

“No…no you shouldn’t have come here, you shouldn’t have gotten involved,” he said as Jimmy looked at him. “I told you…there’s nothing we can do to stop her…it’s too late…She’ll destroy everything in her path until she can see the world at her feet.”

“It’s not too late, Clay,” he then began. “There is still a chance to stop her…there’s still a chance we can end this once and for all and change everything that has happened…we have people who can help us…people I told you to trust…and they will fight for you but we need to make a plan,” he said.

“Stopping Emily doesn’t change anything…” Clay said as Lutheran thought. “I already told you…the city is lost…” he said.

“It’s not about the city anymore, Clay…it’s about helping save lives. Help end something that could be worse,” he said as Clay shook.

“Without the city, there’s nowhere to go. I failed at defending it…and I failed against, Emily,” he said. Jimmy shook as he looked to him. “She brought this down on everyone…all because of me…I won’t let anyone else suffer,” he said.

“Clay…just because we lost the city doesn’t mean it’s over,” he said as Clay shook.

“It is over…regardless if what you say is true, Jim…I can’t do it…it’s time you accept that,” he said as Lutheran ran up to Clay growling.

“What the hell happened to you? Huh?!” he quickly began. “What happen to the Clay Treston I knew? The one who stood up for people…Gave them hope and risked his life for even someone like Jimmy?!" he said as Clay looked away. "You really would give up because you lost? Because you couldn't do it and you won't go through the pain anymore?" he said as Clay stood motionless looking at the ground as Lutheran shook. "You really are that pitiful now? ""If Mya was really here…she would be disappointed in you," he said as Clay nodded. "…you're a coward. And you disgrace her death." He walked past Clay who stood at the door.

Looking down at the ground, Lutheran walked out while Clay's eyes watered. Holding back his tears, Jimmy stood looking at him quietly as he soon nodded his head and spoke.

"Clay…I," he began as Clay shook his head.

"No…no, Jimmy, he's right," he began as Jimmy looked at him. Thinking while he stood, he shook his head as he spoke.

"Clay…I don't understand…wh…why are you giving up?" he asked as Clay stood, looking down.

"Because…I'm scared…I'm scared, Jimmy," he began as Jimmy looked at him. "All my life…I've done my best to survive. I've gone through racism, beatings, abuse…I can honestly say I've gone through it all," he said nodding. "I can handle that. If it's just me, if it's all put onto me, I can handle it…But…never in my life, have I ever faced something like this…like Emily," he stated. "I am…terrified of her Jimmy…She came to this city and she has done everything she could to bring me down…she made me a target and used me," he said shaking his head as he thought. "And because of that…people were hurt…people were killed…Nick was killed," he said as Jimmy shook. "No one has ever done that to me before…no one has ever hurt people around me because of who I am…and I'm scared of that…I'm scared of walking out these doors…I'm scared to go out into the city…I'm scared of people like Emily who are going to do whatever they can to destroy me…and destroy the people that I love…destroy people I don't even know. I can't let anyone else get hurt because of that," he said, looking at him. "And yeah…Lutheran in a way is right…Mya's death was probably my fault…and I am a coward. I am a disgrace to her," he said thinking, as he shook his head. "There is no purpose for someone like me in this world and I wish I would've known it sooner," he stated looking down while Jimmy stood. Thinking for a moment, he soon shook his head as he looked back up at Clay.

"I may not know you very well, Clay" he slowly began. "But you saved my life…and I can't thank you enough…and ever since I met you, I knew you were different. I knew someone like you could make a difference in this world…where someone like me can learn from and follow even," he said as Clay thought. "People believe in you Clay…whether you fail or not. Whatever Emily has done, is done. And yeah you lost Mya, but you can't blame yourself for that," he said as Clay looked at him. "If anything…you did whatever you could to protect her…and I know it hurts…to lose everything…but I also know, that she is still here with you now because she is trying to help you. Just as I am here with you now," he said as Clay looked down while Jimmy spoke. "I don't know if you'll join us Clay…but I know we can make things right…and though we don't have a city, there are still hundreds and even thousands of others out in the world, who could use a Clay Treston right now…and that's a purpose," he said as Clay looked at Jimmy who soon nodded to him and walked past him to head back to camp with Lutheran by his side.

Chapter 22

Back in the east, Commander Barnett stood in his tent. Writing a letter, he soon looked up to see a soldier with curly brown hair and a light beard come in.

"Sir," he began as Barnett shook his head.

"Danny…what can I do for you?" he asked as the soldier, Danny stood.

"More soldiers from Burunian," he said as the Commander looked at four men walking in.

"Only four of you?" he began as they nodded to him. "Why so few?" he asked as one of the soldiers, Randall, spoke.

"We, uh, got split from the others, you could say…" he said as Commander Barnett looked at him while he soon nodded.

"So you traveled all the way from Burunian City?" he asked as Randall nodded. "Um more or less…" he said as the Commander thought and looked away.

"Alright, well, make yourselves at home…it's good to have you," he said.

"Of course. Thank you…sir," he said as he turned away walking with Danny. As they left the tent, another soldier soon came in. Dragging in another, Commander Barnett glanced up to see Adrian in his tent, as he sat up.

"Well, I can't say I'm surprised…" he said as Adrian stood before him brushing off the soldiers hand on her.

"Sorry sir, I found her sneaking around the camp. I thought you should know" he said as Commander Barnett nodded to him as he soon left.

While the soldier walked out, Commander Barnett looked at Adrian before him as he spoke, "What are you doing here, Adrian?" he asked.

"I got tired of staying in Burunian…I wanted to come here to see you…especially with war at your feet," she said as he shook.

"We're not at war" he said.

"Not yet, anyways," she said as he sat back in his chair.

"So you've heard," he said as she nodded.

"You know you can't leave me out of this," she then stated. "If there is anything I want to do, it's to fight beside you."

"You know how dangerous it is Adrian…that's why I kept you back," he said as she spoke.

"I can defend myself," she said.

"And I don't doubt that…" he began. "You're strong, Adrian…but I can't have you here with me, knowing you could get hurt," he said.

"Even if that's the case, you know you can't just throw me to the side and push me away," she said.

"I never meant to, Adrian, things just happened this way and I'm left in a tight situation," he said as she shook. "You know I'm right…that's why I'm sending you back to Burunian," he said standing up as she looked at him.

"Wait, you can't be serious?" she asked as he shook.

"Adrian…I know you understand…this is what's best for you," he said as she looked down thinking, while he walked up to her.

"Look Adrian…I love you…but I can't lose you…not after what happened with your brother. He wouldn't want this," he said as she nodded looking away.

"I know…I know," she said as he soon smiled putting his hands on her arms when all of a sudden, the alarm went off outside.

"What's going on?" Adrian then asked as soldiers began yelling outside.

"Nothing. Stay here," he said walking away as she was left waiting inside the tent.

Walking out, Commander Barnett looked around at soldiers running through the camp. Seeing them pass him, he soon stopped a soldier and spoke.

"Why was the alarm set off?" he asked.

"Up ahead, wounded scout soldiers," he said as Commander looked to where he was pointing as he caught up with the rest of the soldiers.

With all of them gathering in the open field, two soldiers dressed in HOPE uniform helped one another walk while they were covered in blood. Eventually reaching the camp, soldiers surrounded them as Commander Barnett worked his way to them. Seeing them on the ground, catching their breath, he knelt down and spoke.

"What happened here?" he asked the soldiers.

"We…we were in the woods," One said in pain as he shook his head. "Scouting the area…when IOBs appeared…They're coming…lurking in the woods," he said as Commander Barnett quickly stood up, looking toward the woods. Seeing nothing around them, he soon turned to the soldiers beside him as he quickly spoke.

"Get them to a medic…" he said as a group of soldier began to help them. "The rest of you," he then said to those around him. "Get to your posts and gear up! This isn't a drill," he yelled out as the soldiers all soon went to do so, yelling amongst each other, while he stood looking at the woods, preparing himself for a fight with the IOBs.

Chapter 23

Walking back through the camp, Commander Barnett ordered troops around as Adrian stepped out of the tent. Looking around, she saw her father as she walked up to him and spoke.

"What's happening?" she asked as the Commander ordered a soldier away.

"IOBs," he said looking at her. "They've been spotted not far from here," he said as he thought it over. "I don't know how it's possible…but it is. And we're in danger," he said as she shook.

"I can fight," she then stated as he quickly shook his head.

"No. I told you…you need to head back to Burunian but until then, you need to head towards the woods. It's the safest place for cover," he said as she shook.

"No. I can't do that…if there is any way for me to be useful, then I should be here," she said as he shook. "It's not about being useful Adrian…it's about surviving," he then said as Adrian looked to him as all of a sudden above them, a shadow was cast over them.

Covering the skies above the camp, the Commander along with Adrian and the rest of the soldiers looked up as the IOB warship could be seen, slowly hovering in the air over them.

Commander Barnett watched the ship hover in place while it began opening up ports. With metal clanking over them, Commander Barnett looked towards Adrian as he quickly shook his head and pushed her away.

Nearly falling as she stepped back, she could only look at her father yelling out, as everything else blanked out.

"Hurry! Go now!!!" he yelled out to her as she slowly shook her head in confusion when all of a sudden, the ports on the IOB ship began shooting out pods towards the ground. Crashing down all around the camp, Adrian felt the ground shake while soldiers yelled all around her. Looking around slowly, she soon looked back at her father yelling out before her when all of a sudden, a pod crashed down, landing on top of him. Adrian moved her face away as dirt shot out all around her.

Finding herself lying on the ground, her hands were shaking as she realized she had her father's blood all over her face and hands.

Looking at the blood, she was in shock. She looked back towards the pod before her as it opened up and began releasing IOBs.

Quickly screeching, they quickly began attacking as soldiers ran out before Adrian to hold off the creatures.

With soldier after soldier being struck down while she continued to lay still, she could only look at the pod on top of her father that was just before her as she tried to process what had happened.

Trying to understand, she looked towards a soldier before her as he was holding off an IOB. With a blade being stabbed into him, he looked back to Adrian as everything around her was muted. Looking at him, she faintly heard him shaking his head and yelling as he told her to get out of there.

Focusing on him, blood began running from his mouth as the IOB dropped him before her. Looking at her, the IOB soon lunged towards Adrian as she quickly pulled out a blade and plunged it into its head.

Holding it back as she pulled out the blade, it soon stepped back screeching as she quickly looked around her and got up.

Getting to her feet, she quickly stood up and began turning towards the woods as she realized there was nothing she could do.

Beginning to run, as she quickly looked before her, she found herself sprinting.

With explosions and soldiers being blasted around, she could only keep herself running towards the woods as tears began to run from her eyes.

Chapter 24

Two years ago

Standing in the city, Holly stood over a fallen building as below her, a crushed street full of blood and bodies could be seen in the city as she slowly thought something over.

Shaking her head, she looked over at Emily beside her, while they slowly spoke.

"This isn't right," Holly slowly began. "All these people we've killed…just so we could see another day," she said.

"It's not your fault, Holly…we care about our own lives too much. That's just the human side within us," she said.

"But you knew these people…" she then began looking at Emily. "This was your city…a city of people you've helped to survive," she said.

"Even if there were another way, Holly…they wouldn't have made it out of this alive," she said while she thought. "The only thing we can do now is move on. Do whatever we can to protect ourselves…even if it means making sacrifices," she said as Holly thought.

While she did, a sudden ringing noise went off in the air as Holly and Emily looked up for a moment and spoke.

"They're calling us. We should head back," Emily stated as she turned away while Holly stood. "Aren't you coming?" she then asked as Holly slowly shook.

"I…I can't," she then said as Emily shook.

"What do you mean?" she then asked as Holly looked at her.

"I can't go back Emily…I can't. I can't do this anymore," she then said.

"No offense, Holly, but if we don't make it back, they'll easily kill us within a matter of minutes," she said as Holly shook her head.

"Would they? Would they really do it?" she then asked as Emily stood. "I mean think about it…we were created to be superior to them…they made us better than they could ever dream of being. If that's true…then why should we follow them? Why should we fear them?" she then asked.

"I wouldn't tempt them, Holly…" she said.

"Maybe…but one of these days, Emily…we're going to escape. We're going to run…and start a new life and forget all this ever happened," she said as Emily thought. "No matter where we go…we'll find something better. Anything is better than here," she then said as her and Emily stood while all of a sudden an IOB ship appeared over them. Looking up, Emily slowly shook her head as she spoke.

"Well…maybe next time," she said while they were taken by the IOBs.

Present Day

Standing in her tent, Holly slowly packed her gear as the grey clouds hung all around her. Eventually lifting up her bag and putting it over her shoulder, she turned around and she began to walk. Working her way through people, she held her head down under her helmet. Heading south, she planned on running away from the city to get away from Emily. With no other choice, she couldn't help but feel the thoughts rushing into her head while she soon walked out.

Feeling herself trying to hold back her emotions, she soon stopped with people walking all around her. Breathing heavily as she stood in the street, she slung off her bag and put it on the ground. Shaking her head, she took off her helmet and she held it in her hand.

Looking at the blue helmet with the IOBCF logo across it, she felt hurt that she wanted to leave.

While she stood in the street, she soon looked beside her. Not far from her, she looked into the tent of Liz. Liz was slowly helping the patients she had in there while Holly stood looking at her.

Seeing her, Holly felt herself smile a bit as she saw Liz there before her. Thinking, she knew deep down inside she couldn't leave, as she knew she had to protect her.

As Holly stood, another woman in an IOBCF uniform and short red-brown hair stood beside her, and slowly spoke.

"She your girlfriend?" she asked as Holly soon looked at her, beside her and nodded. "Um yeah…how did you know?" she asked as the woman stood.

"Right, sorry, my name is Kate. Ever since Emily came here to the city and joined IOBCF, she's been a good friend of mine," she said.

"Oh…I didn't know," she then said as Kate nodded.

"Emily said you were the 'officer' under her and that's how I know you…she said I could find you around your girlfriend, Liz," she said.

"Right," Holly said as she thought. "…you new here, Kate?" she then asked as Kate nodded.

"Yeah! Well, kind of…Emily just let me into her ranks…I've lived in this city my whole life and only now am I ready to actually fight," she said as Holly listened.

"Well…when you have something to fight for, it helps" she said.

"Yeah maybe," she said as Holly nodded and looked back at Liz working.

"So, I take it you fight for her?" Kate then asked.

"I wish…" she said as Kate looked at her. "It's complicated," she then said to her. "We've grown distant from each other lately…I guess I haven't had my priorities straight," she said as Kate shook.

"Do any of us?" she then chuckled as Holly nodded. "Yeah…" she simply said thinking, while Kate looked.

"You know…sometimes to get the person you love to trust you…you simply just have to fight for them," she then said as Holly looked at her. "I hope that helps…nice to meet you, Holly," she then said walking away as Holly stood.

Taking a moment to think, she eventually picked up her bag off the ground.

Standing back up, she put her helmet back on. Thinking while she looked at Liz across the street, she soon felt herself turn away and head back towards her tent.

Chapter 25

Walking back to camp, Jimmy held his head low while Lutheran walked beside him. As they walked through the tall grass around them, they looked up. Seeing a stream of smoke rising up from the camp in the distance, Jimmy slowly spoke.

"What's going on?" he began as Lutheran looked.

"Something's not right…C'mon!" Lutheran grunted as he soon went running towards the camp.

"Wait, Lutheran!" he began as he looked back up at the smoke rising up. "Wait up!" he said running after him as they went to check out the scene.

Running ahead, Lutheran immediately dashed through the camp while Jimmy stood catching his breath. While he did, he looked around him at tents destroyed in the camp and the vehicles lit on fire. Looking around, soldiers sat around the camp covered in dirt as Jimmy looked to them.

"Wha…what happened here?!" Jimmy asked a soldier.

"IOBs…they came in and attacked us…like they knew we were here," he said as Jimmy shook while Lutheran walked back towards him. Dropping a device down at his feet, Lutheran slowly shook his head as he looked at it.

"It's some kind of tracking device…It must have been what led the IOBs here," he said as Jimmy shook.

"No…No this can't be right," he said as he looked around at the soldiers. "Where's Commander Barnett?" he asked as the soldiers looked around at each other. Quiet while Jimmy stood amongst them, soldiers soon walked up to him as they dropped Emily's soldier Randall before them.

"I noticed this soldier trying to head out of camp," the soldier, Danny, stated as he looked at him. "I recognize he's one of the four soldiers that I brought in…He could have been the one that planted that device," he said as Lutheran stood looking at Randall.

"They were IOB-induced people," Lutheran began. "…people from the city" he said.

"Well you're not wrong, dog" Randall then began. "Emily told us about a group of soldiers stationed outside the city…apparently, she had a suspicion that you would come to try and stop her…but we didn't want it to happen. So we brought in the IOBs…and they killed your so-called commander," he said. Jimmy ran up to him, grabbing him by the collar as he yelled at his face.

"No! You're lying!" he screamed as the man laughed.

"Think what you will, kid, but he was blown to bits…ask the men and women around you," he said as Jimmy let him go and began to look around.

"Is it true? Is he dead?" he asked as the soldiers looked down. "Is he dead?!" he yelled out as Danny slowly nodded his head.

"He is, Jim…I saw it…he's gone," he said as Jimmy felt his body go all of a sudden numb. Feeling himself choke up, he soon knelt down before the bald soldier as he shook his head and spoke.

"No, no…no, no, no, no, this can't be right!" he said to himself.

"Oh, it is, kid…and it's only a matter of time before you share the same fate" he said as Jimmy slowly then looked up at him before he quickly threw a punch at the man while soldiers all went towards him.

Eventually, pulling Jimmy off Randall, they held him back as he was struggling to hurt the man.

Lutheran quickly looked to him as he spoke, "Jimmy! Get a hold of yourself!" he said. Jimmy shook and looked at him.

Pushing the soldiers off him, he shook his head as he breathed heavily looking at Lutheran.

"Get a hold of myself? Are you serious Lutheran?" he then began. "Look around you! The commander is dead! Emily knows we're here and her war is about to begin…what are we possibly supposed to do now?" he asked as Lutheran shook.

"What we have to do," he said as Jimmy shook.

"There's no commander, Lutheran, we don't have anyone," he stated as Lutheran shook. "Then someone else will have to take command!" he grunted as Jimmy shook.

"It doesn't work like that Lutheran…we…We can't just go on ahead!" he said as Lutheran shook.

"Then what do you suggest?! To ignore the war that is happening outside our doors? To let Emily go while we cower in fear? We still have a mission to take care of…even if it means you taking command," he said as Jimmy soon shook his head.

"Look, Lutheran…I know what's at stake, but I can't lead these men. I can't do that," he said.

"If you won't lead these men, then who will?!" Lutheran then asked as all of a sudden a voice spoke.

"I will," the voice said as soldiers looked amongst each other to see Adrian walking through them. Seeing her step forward, Jimmy could only look at her slowly as he spoke.

"Adrian…" he began.

Lutheran looked at her while he squinted his eyes. "What is your name?" he asked.

"Adrian Barnett. Daughter of Commander Barnett" she said.

Lutheran nodded.

"If anyone should lead these soldiers, it should be me" she said.

"Wait, Adrian," Jimmy said.

"Jimmy…" she said as he stopped. "Don't…this is something I have to do," she said while he stood thinking to himself.

"And can you fight?" Lutheran asked as Adrian looked at him.

"More than anyone here could," she said as Lutheran stood thinking while Jimmy spoke to him. "Lutheran…no…" he said as Lutheran looked at him. "You can't do this to her…"

"Unless you're willing to be commander, Jim…I think we're out of options," he said as Jimmy stood thinking while he looked back to Adrian.

"There's a lot here going on…things you may not understand. If you do this, we'll need your full support on whatever plans we may develop," Lutheran said as she nodded.

"Then you'll have it," she stated as he nodded at her.

"Then it's yours," he said as Jimmy looked at him. "If you believe you can lead these men and women into battle…then you will lead us," he said.

"My father believed in the soldiers around him…If he agreed to fight then it means it was for some kind of purpose…a purpose of having faith in the soldiers around him and the trust to have them stand by him," she said looking at Jimmy as he and the rest of the soldiers stood all around while Adrian turned to him. "I know everyone here…they're all my friends…and we have a purpose…my father believed we could be so much more in this world…when he created HOPE, it wasn't for the purpose of just fighting IOBs…but the purpose to stop all kinds of evil in this world…and if that's what we have to do…and if that's why we're here…then I know every one of them here, will support me," she said to the men and women around her as Danny then nodded at her. Looking at them all, she soon turned back to Lutheran as she spoke.

"My father trusted you…and if he did, then so do I," she stated as he nodded. "We'll fight with you," she said as Lutheran looked at her and stood up.

"Then it's settled…" he said to her as she nodded and turned to her soldiers. Looking at each and every one of them, she knew she had their support with her, while she turned to Randall.

"Throw him in the woods…he can rot there," she said as soldiers picked him while he chuckled and dragged him away.

Heading away from the scene, Adrian slowly then turned to Lutheran and Jimmy as she nodded her head.

"Thank you for this," she said as Lutheran simply nodded. Standing, she soon looked at Jimmy who could only look at her shaking his head while he spoke.

"Adrian…this isn't right," he then began. "What lies in the city…is something we've never ever seen before," he said as Adrian shook.

"Then I will be strong like my father taught me," she stated as he stood. "And I'll use it for war," she said as she walked away.

Chapter 26

In the night, Clay stood outside his cabin as he looked around the woods, taking in a breath of air. With his mind blank, he held himself together, he knew war was about to begin in the city. He couldn't help but wonder what was going to happen. As he stood thinking to himself, he soon looked over as in the woods, he heard something large move around, it brushed against the twigs and bushes.

With trees creaking, he stood looking in their direction, while it soon went quiet. Slowly he then went walking towards the sounds.

Going over to the bushes in the woods, he stood looking around as he continued to hear something moving ahead, when it then took off.

Seeing something large run off, he quickly began to follow it as he ran to see what it was.

Everything seemed to appear around him vastly different, he ran in the dark squinting his eyes with the wind hitting him. Trying to keep his eyes on a shadowy figure that was getting out of his sight, he continued to press forward.

Running around endless trees and obstacles, Clay soon stopped as he looked around, catching his breath and he couldn't see anything. Looking around, he tried to think where the figure went as all of a sudden, ahead, a bright light began to shine while he covered his eyes with one hand and looked in the sky.

Slowly seeing it, he began to head towards it as he worked around some more bushes and trees and he soon noticed it above him, he was standing under what looked to be a massive ship. Seeing it, he quickly noticed it was an IOB ship as he began stepping back while something came up behind and knocked him unconscious.

Chapter 27

Standing in her tent, Liz sat quietly as she held a picture of her and Holly in her hand. Crumpled up, it was fading, she soon put her hand over it while she closed her eyes thinking. Holding back her emotions, Holly walked in as Liz slowly turned to her.

"Holly…" she began as Holly stood looking at her. "What are you doing here?" she asked as Holly quickly hugged and kissed Liz. Putting her hands around her face, Liz shook herself away from Holly and spoke.

"No…No, I can't," she began as Holly looked at her.

"Liz…I" she began.

"No, you don't get it, Holly…you can't just walk in here and expect everything to be just okay," she stated while Holly shook.

"But Liz…I came here to apologize to you," she said.

"Apologize…after all this time…do you honestly think an apology can fix what you've done?" Liz said as she looked at Holly.

"I…I know what I've done isn't right, Liz…but you have to understand, I only did it to protect you." she stated while Liz shook.

"Protect me?! You destroyed most of Sector A! You nearly killed Clay and you've taken yourself to stand beside Emily, like some kind of pawn," she said.

"I had no choice…If I didn't do what I needed to do, then Emily would have hurt you…I just wanted to take care of you" she said slowly, with her voice trembling.

"Well good, good, thanks for protecting me Holl…because you've done a great job caring for me," she said looking at her while Holly thought.

"Liz…I care about you…I love you…" she began looking at her, "I love you and I would do anything for you…just…please…just…tell me what I need to do to make up for it," she said as Liz stood looking down while she spoke.

"You wanna know what you can do to help me, Holl? Do you want to know?" she said looking back at her. "Leave…leave Emily and be with me…be with me for a change and actually show you that you care," she said as Holly looked down shaking her head.

"It's not that simple, Liz," she began as Liz shook.

"No, Holl…you're wrong…it is that simple…that's all you have to do!" she shouted as Holly looked at her.

"I…I can't…if I do, Emily will know…she'll know and she'll kill you…and she'll kill me the first chance she gets," she said as Liz slowly shook her head.

"I'm sorry, Holl…but it's the only way," she said as Holly looked at her, trying to speak when all of a sudden a voice spoke from behind.

"Well…isn't this all a surprise?" the voice said as Holly turned around to see Emily. "No…" she said as Emily walked up to stand before them.

"You know, Holly…" she slowly began walking in. "when I told you to end things with Liz…I meant every word."

"No, stay away," Holly slowly began. "This is none of your business, Em," she said as Emily shook her head.

"I'm afraid it's too late for that, Holly…and it is my business now…breaking my rules has consequences…you of all people should know that," she said as Holly shook.

"No…no, it doesn't have to be this way," she said as Emily smiled and began walking around.

"Well…I suppose there is another way," she then said as Holly looked at her. "I'll tell you what Holly…I'll give you one last chance…you make your girlfriend, Liz here, join us…and we can just forget about all this…forget what has happened and move on," she stated while Holly thought and looked up at Emily. "If she does that…then I'll leave," she said as she then looked up at Liz and spoke. "So what do you think, Liz…if you really love Holly, would you end her suffering?" she asked walking to her. "Would you put away your hate for me and join my army so you could be with Holly?" she asked while Holly soon turned behind her to Liz who stood thinking. Thinking what to do, she froze looking into Holly's eyes for a moment.

With her feelings overwhelming her, she soon held her head high as she looked back at Emily before her. Looking into her face, she soon shook her head as she spoke.

"I'll never join you," she said as Holly began whimpering, while Emily stood nodding her head.

"You're strong, Liz…I'll give you that," she then began walking away while Liz looked to Holly falling to her knees.

While she did, Emily held her back to them as she took a breath of air in and spoke.

"Well, if that's the case…" she then said as beside her, Kate came walking in as Holly looked at her and saw a rifle in her hand. "Then I have no choice but to kill you," she said as Holly quickly looked at her and stood.

"No…no, no, no, no! We agreed! We agreed she wouldn't get hurt if I helped you!" she cried.

"And I said I wouldn't, unless I needed to…if she's not willing to join, then she must die."

"No, no, no, Emily, please…you can't do this…I've always helped you and did what I had to do for you," she said as Emily shook.

"And now I'm helping you…so move aside Holly…or you will also share the same fate," she said looking back at Liz while Holly thought.

"So Liz…shall we decide again to take my offer?" she said as Liz shook her head nervously as tears ran down her face while she looked at Holly, who turned to her.

"Like I said…I'll never follow you," she slowly said looking back to Emily. "You bitch," she stated as Emily soon nodded her head and looked to Kate.

"So be it…Kate, kill her," she said as Kate lifted the gun up as Holly yelled out.

"No, wait!" she said as Emily looked at her. "I'll do it…" she said looking at Liz as she spoke. "…I'll finish her," she said to Emily who soon smiled and looked at Kate who put down the gun and handed it to Holly. Looking down at the gun and Holly slowly took it as she held it shaking, while she thought.

With it in her hands, she soon turned back to Liz as she held back her tears.

Standing still, Liz held her head still high while she looked at Holly. Knowing what could happen, she slowly began to speak as Holly pointed the gun at her.

"If you have to do it…then do it," she began as Holly shook, breathing heavily. "Otherwise…if you love me…if you truly love me," she said as Holly looked to her. "You prove it now," she said with her voice cracking as Holly breathed heavily, trying to think what to do.

"Do it, Holly…kill her," she said as Holly continued to hold up the gun before her as she shook, thinking and thinking as her mind ran wild. Trying to come up with something, she could only look into Liz's eyes, feeling herself tighten up as she eventually began lowering the gun.

Watching her, Liz soon let out a sigh of relief as Holly lowered the gun down beside her and dropped it to the ground while Emily looked to her and spoke.

"You sicken me…" she began as Kate walked up beside her and took the gun. Quickly then, Holly ran up to Liz and hugged her as they held each other tightly.

"I'm so sorry Liz…please forgive me," she said as Liz could only shake her head while she spoke. "I already have, Holl. I already have," she said lightly as Emily slowly walked up to them.

"I saved your life…" she began slowly as Holly looked at her. "And yet you betray me…do I mean nothing to you?" she began as Holly looked at Liz.

"Now that I see who you really are. You were never anything to me," she said as Emily cringed looking at her. Standing before them, she then quickly pulled out a knife beside her and cut across Holly's arm as she lunged. With Liz kneeling down with her as blood ran from her arm, Emily stepped back as she looked at Kate who soon walked up to them and knelt down dipping her hands in the blood on the ground.

As she did, the Liz and Holly looked at Emily as she slowly cleaned off her blade with a rag and spoke, "If you won't join with me as my creature…" she began while Kate stood by, licking the blood off her fingers. "Then I'll have another," she said as Holly slowly shook.

"No," she said as Kate's eyes began to turn black while Emily turned away. Not turning to look at the two of them on the floor, Emily shook leaving the tent, while Kate followed behind.

With them leaving, Liz shook her head, checking out Holly's arm as it bled while she spoke. "Are you alright?!" she asked as Holly nodded.

"I'm fine…" she said as she looked out the tent. "What do we do now?" she asked as Liz shook.

"We have to go…there's people outside the city that can help us…we just need to get out of here quickly!" she said as Holly nodded.

While they spoke, outside the tent a soldier walked up to Emily as he quickly spoke beside her. "You asked for me, Commander?" he said as Emily nodded while Liz and Holly prepared to go.

"Kill them," she said. The soldier nodded and worked his way to the tent as Liz and Holly stood. Turning back, they noticed the man standing in the entrance as he slowly pulled out a gun on them.

With him pointing it to them, their eyes soon opened up as gunshots could be heard flashing into the tent while Emily stood smiling with her army gathering in the streets.

Chapter 28

Within the night, Jimmy could be seen standing in his tent as he looked down at a map of Hollandview city on the table. Leaning over it, he let out a deep breath as he thought while Adrian soon appeared outside his tent.

"Jimmy…" she began as Jimmy looked to her as he stood up straight. "Adrian…hey…you alright?" he then asked. Adrian nodded.

"Um yeah, I'm fine," she said looking at him. "Is it alright if I come in?" she asked as he slowly then nodded.

Walking towards him, she slowly walked in and stood beside him. Looking down at the map, she nodded her head as she cleared her throat and spoke.

"Um, Lutheran told me your plan before he set out into the city…I think it's pretty good," she said as he shook.

"Yeah it's uh…it's hard to say, I guess" he then said looking back towards the map. "Holding off Emily's forces and the IOBs at the same time is going to be difficult," he said. "But…Hopefully, Lutheran can distract her long enough until we can do what we need to do," he said as she shook.

"So that would explain why he took a small task force with him," she said as Jimmy nodded.

"Yeah, uh, once we're able to manage the IOBs, we'll be able to rendezvous with him. Hopefully he can hang in there long enough…if not, then I'm not really sure what we'll do," he said.

"Well…If we stick the plan, I'm sure we'll be alright," she said as he looked down nodding while she stood. Thinking to herself as she looked at him, she soon shook her head while she took a deep breath and spoke.

"Hey um…Look, Jim…I…I wanted to apologize for today," she said. "I should have told you I was taking command and…I'm sorry you had to find out this way," she said.

"Uh no, it's alright, Adrian. You didn't do anything wrong. If anything, I was being a jerk. It should be me apologizing," he said as she looked at him. "I was wrong to think you couldn't do this, and I was wrong to try and hold you back. I guess I've lost a lot of people close to me since the invasion began and ever since your dad brought me into HOPE, you and him have been the closest thing to a family that I've ever had," he said as she stood. "I wanted to look out for you because I didn't want to lose you. I thought with what happened with your dad might affect your decisions as commander," he said. "But I realized then that I was wrong there too, because I know you and I know you're strong and smart and have something in you that no one else has," he said as she nodded. "And I know that in the end, you'll do what's right…" he said as she nodded her head.

Thinking for a moment, she stood while she looked down and slowly began to speak. "Well, you don't have to worry about me, Jimmy" she then said as she looked to him. "At first…I guess I didn't really know why I chose to be commander…and in a way, you were right…there was nothing more I wanted to do then get revenge…to make Emily pay for what she did to my father…" she stated. "But I realize if I did, it would make me no different than her…and everything I was taught would be for nothing," she said as Jimmy thought. "My father prepared me for life…but he didn't prepare me for this. And I didn't know what to do…I didn't realize what I was doing…until I started thinking about my brother," she said as Jimmy looked at her shaking her head.

"My brother was a strong man…and before he died, he told me once, that whenever an opportunity rose up, to always take it. No matter how hard it may be or how impossible it may seem. That I had to push myself to be the best person that I could be and to prove to everyone around me that I deserve this," she said as Jimmy looked to her. "And that's when I realized I needed to do this Jim…not just for myself…but for him. Because I knew he would want me to do this. And I knew there was nothing more that I wanted than to prove who I am in this world and what I really can do," she said as Jimmy nodded. Thinking to himself, he soon shook his head and spoke.

"Then we'll do this together. And I'll do whatever I can to help," he said holding her hand as she nodded.

“That’s all that I needed to know,” she said then hugging him as he stood. Standing as he held her, he soon smiled.

“So…does this mean we’re ok again?” he asked as she smiled. “Oh, more than ok,” she said as they stood together.

Eventually letting him go, she stood hearing soldiers call out for her while she spoke.

“Right…I forgot I made plans moving camp closer to the city,” she said as he nodded. “I figured it would be easier to know when the fighting would begin…so I guess I’ll see you in the morning?” she asked.

“I’ll see you then,” he said as she then nodded and began to work her way out.

As she did, Jimmy stood letting out a smile as he nodded thinking. Looking back down at the map though, he knew their focus needed to be on Emily while he stood thinking. With their plan in sight, it was only a matter of time before they would face Emily and the IOBs, and he stood wondering if his plan would work.

Chapter 29

Breathing slowly, Clay began to wake up. With his eyes slowly moving, he soon looked around him with his vision coming back to him. With bubbles floating around him, he noticed a mask over his voice. Trying to speak, he looked around him and found that he was stuck inside a chamber of yellow water.

Lying on his back, he soon looked around thinking. Putting his hand on the glass, he began pounding on it while he noticed himself in some kind of dark room.

Pounding the glass, he soon noticed an alarm set off beside him. Looking at it, he watched as the water around him began to drain.

Watching the water drain, he looked above him to the top of the chamber as it opened up. Sitting up, he quickly felt the blood rushing to his head while he waited for a moment and took off the mask around his face.

Pulling it off, he slowly let out a small breath of air as he quivered. Taking a moment to breath, he soon looked at his body while he was in his drawers and had cords hooked to him. Ripping them off him then, he soon sat up looking around so he had a better view of where he was.

With black metallic plating all around the room, small lights dangled on the ceiling with lines of cords everywhere as if someone placed them there. Finding himself in this weird complex, he soon heard a voice speak.

"Good of you to wake up," a male voice began clearly as Clay looked around for the voice speaking. As he did, he soon cleared his throat and spoke.

"Where…where am I?" he began as the voice spoke.

"You're safe. I managed to capture you to help make sense of things," he said as Clay shook. "I don't understand…what do you want with me?" he asked as the voice continued.

"Well, for starters…I healed your body already," he said as Clay thought as he looked at the wounds on his stomach while they were completely healed. "It was strange considering you didn't heal as fast as you usually do. You must have been very stressed," he said as Clay looked around and spoke.

"How? How do you know that?" he asked as the voice chuckled.

"Well, I should know considering I'm the one that did this to you," he then said while Clay then looked beside him as a door opened up. An IOB soon walked out. Seeing the creature before him, Clay looked around for his suit and gear and found them placed on the floor ahead as he slowly looked back to the IOB speaking,

"There is no need for that…you have nothing to fear. Clay," he said as Clay sat in the chamber, confused. "How are you able to talk? How do you know me?" he then asked as the IOB shook.

"Well, like I said I'm the one who did this to you," he began as he walked closer to Clay who sat looking at the creature who tossed him a blanket as he wrapped it around him. "And talking-wise, well I had a friend. He was human…he was a good man until the others got to him…after that I decided to incorporate his voice through this collar I created," he said pointing to it, as Clay shook as he spoke.

"But IOBs don't talk," he said as the IOB shook.

"True. Until now," he said continuing walking around the room. Clay squinted as he looked to him and thought as he shook.

"You mentioned that you 'did this to me'," Clay began as the IOB turned to him. "What do you mean by that?" he said as the IOB nodded as he shook.

"Do you remember Clay…the time you were captured by an IOB?" he said as Clay nodded.

"Um yeah…but it was a long time ago. I don't…really know what happened though," he said as the IOB spoke.

"Well to tell you the truth…I was there," he then said as Clay looked to him. "What? But how?" he asked as the IOB spoke.

"Well considering I was there when you were captured…I should know," he said as Clay shook while the IOB then nodded. "You were trapped Clay…stuck on a table where they had you tied up. You were to be cut open but instead, I injected a fluid into you…fluid containing our DNA that made you who you are today," he said as Clay thought. "From there, I loosened your restraints and escaped with you….afterwards…I dropped you back off in the city leaving the suit you wear now with you in hope you could use it to fight back," he said as Clay shook his head while he spoke.

"So you…saved me," he said, looking up to see the creature chuckle. "You see, not all IOBs are bad," he said while Clay shook.

"But wait…why? Why did you save me?" he asked as the IOB stood.

"Well, there were many reasons before I chose you," he then began. "Before even you came along, I watched as we came to earth. I was in charge of the research division within our ranks. I experimented on things to study life forms and technology of different planets we travel to…but when I came here to Earth…I watched my people become more ruthless than they have ever been before…they struck down knowing what was here and they wanted to take it all away…they wanted to take control over it…show how superior they were to humans," he said as Clay listened. "It wasn't until I watched my people…begin to kill humans. Killing them in such ways that…that shouldn't have been done. It wasn't until one day I met with a human trapped here while others worked on him. He was experimented on numerous times…but yet he kept going…I remember talking to him every day…how he told me about himself…about love, sharing, caring…how he looked forward to what was to come next as he knew he would eventually be happy," he said.

"What happened to him?" he then asked as the IOB thought to himself for a moment and looked down before speaking.

"They…uh…they killed him. As they said I was becoming like him…I was becoming human in a way…and so they disposed of him…and lowered my ranks to a simpler tool holder, you could say…and that was when I told myself enough was enough…and what we did in life…was actually wrong and I needed to do something," he said as Clay looked at him taking a breath in.

"So you chose me," he said as the IOB nodded.

"I saw your thoughts…and I read your mind when the IOBs worked on you. I saw your pain…your suffering…how you lost your wife Mya that day…and you came fighting with such anger and hate towards us…I was ashamed to call myself a proud IOB after that," he said as Clay closed his eyes and shook his head thinking. "I wanted to then to somehow create someone who could fight…someone who could change what we' have done. Someone who could stop my people and lead this world back to what it was," he said as Clay looked back at him as he spoke.

"But what if I didn't want this? What if I didn't want this suffering…Did you ever consider that?" he said as the IOB shook.

"I knew you didn't. I knew you wanted to be with Mya but someone had to do it. And you with your will to fight and the care you have for people, I knew you could," he said as Clay shook.

"Well, I think you chose the wrong person because do you have any idea what I've done?" he then began. "All the pain and suffering I've caused. I destroyed a city. I put people's lives in danger. I…I did the complete opposite of what you wanted me to do," he said as the IOB nodded.

"I know. I have watched you this whole time. After I left the ship, I broke away from my people. I betrayed them. From there, I could only watch, hoping you could save your earth one day…and ever since, I watched you fight. I watched you make friends including your friend, Lutheran, who I helped escape from his prison as well," he said as Clay looked at him. "I know you can do it…and you know you can do it…you just happened to fall the wrong way," he said.

"I've fallen further than just the wrong way," he then said as the IOB nodded before him as he spoke. "True and it's why I'm talking to you now, Clay…it's why we are even here meeting together," he said as Clay simply let out a deep breathe, shaking his head, thinking. "I know you're angry, Clay…but the time now has come where I am helping you…and the time now to tell you what you have to do to stop Emily," he said as Clay looked at him.

"There is no stopping her…she can't be stopped anymore," he said as the IOB shook his head.

"That's where you're wrong," he said as Clay looked at him with confusion. "You see, I created Emily. I infused the cells in her long ago when she broke out, indeed it was the time I didn't see the truth but my people forced me to make her into what she is now…a weapon…a weapon against mankind…but I knew from that moment that she would be deadly…and I created flaws in the cells that she was given," he said as Clay spoke.

"So, you're saying she can be stopped?" he said as he looked at him and smiled, as he nodded.

"That is, of course, if you're willing to believe you can?" he said as Clay soon thought sitting within the chamber as he shook not saying a word as the IOB nodded. "Well, I'm sure after I convince you, you will believe me…until then you'll just have to trust me," he said holding out his clawed hand to Clay. Clay looked at it and the IOB looked at him. Slowly, taking his hand, the IOB helped him stand up from the chamber as Clay stood.

"So you have a name?" Clay asked the IOB who shook.

"On our planet…we don't really have names…" He began, "but I have been fond of the name, Jake," he said as Clay nodded. "Anyway, now that we have fully met each other, there's something else I have to show you before we begin," he said as Clay shook.

"What is that?" he said as the IOB, Jake, soon pulled out a syringe before him and quickly then stuck it into his neck.

"What…what did you do to me…?" Clay began as he slowly felt his body giving out as he couldn't stand. Falling down onto the ground, his vision began getting blurry and he felt weak while he laid on the ground.

Looking up he could see Jake. Jake slowly walked up to him while he spoke. "Something to help you find your way back," he said as Clay soon closed his eyes.

Part 3

Chapter 30

Two years ago

Walking through the streets, Emily looked around with her eyes wide open. Up ahead, it could be seen that a large group of people were standing within the streets as they lived in the destroyed city.

Looking around, Emily slowly walked. With people gathered beside a car, a woman with short blonde hair soon looked towards her and spoke.

"Oh my god...Emily," she said as she quickly then ran towards her as Emily stopped. Looking up slowly, the women quickly went up to her hugging her, as Emily stood still.

"Emily...it's you...it's really you," she said as Emily slowly shook and spoke. "Cassidy..." she said as the woman nodded.

"Yes...it's me," she said with tears running down her face. "Your sister," she said as Emily nodded with tears running down her face as she chuckled.

"Where...where have you been?" her sister, Cassidy, then began as Emily shook. "It's been like...a year since you've been gone. You disappeared. I searched every day for you," she said as Emily thought while she spoke.

"It's...it's hard to explain," she said looking down while Cassidy nodded.

"Well, it's ok...it's ok. Now that you're here," she said as Emily nodded and she looked up to see the people all around the streets.

"Are these all the people left?" she began as Cassidy nodded, looking back.

"It is. We've been trying to fight off the IOBs for some time now...but we haven't been able to get away from the city," she said as Emily shook.

"You shouldn't be here...you should've still left," she then said as Cassidy looked to her.

"Well I...I would have left but I wanted to stay behind...I wanted to find you," she said as Emily then shook as she moved hair away from her sister's face. Thinking, she soon closed her eyes as she slowly then spoke.

"You should have left," she whispered as her sister shook. "What?" she asked as Emily then opened her eyes back up.

"You should have left," she said forming her hand into a blade as she then stuck it into her sister. Gasping, Emily held the blade in while her sister spoke.

"Em...Emily...wh...why?" she asked gasping for air while blood ran from her mouth.

"You should have left...they're coming for you...and I can't let them kill you," she said as she then pulled the blade out while her sister fell.

"Em...Emily..." she said as Emily looked down shaking her head while tears began to roll down her cheek.

"I'm sorry Cassidy...but it was me or them...and I have to do what I need to do to survive," she said as her sister struggled while Emily looked up ahead. When the people in the street looked at what had happened, they began stepping back.

Running away to tell the others, Emily soon shook her head as she looked back down to her sister. "I'm sorry..." she said, as Holly in her IOB form soon flew over her as she let out a screech while she engulfed the street in fire.

Chapter 31

Slowly, the shower of ashes fell from the sky. Down below, thousands of troops in the city of Hollandview stood looking ahead in the distance while the ash fell. With their eyes glowing black and their veins beginning to glow purple, they stood together in IOBCF uniforms while Emily could be seen in the back.

Appearing behind her army, she walked ahead on the rooftop of the building. Walking past IOBCF soldiers lined up beside her, she walked to the edge and stopped. Looking on, she simply smiled at the scene of troops across the city. With soldiers in the streets and rooftops, she stood waiting and looking ahead while they lay in wait in the city.

Thinking to herself then, she soon looked beside her where she could see Randall walking up to her. Seeing him, she slowly looked back towards the city as she spoke.

“I have to say, Randall…it took you long enough to get back here,” she began as Randall shook.

“Sorry…I got tied up with some HOPE troops…luckily I managed to break free before it got ugly,” he said.

“Good…then all is going according to plan,” she then stated as he spoke.

“More or less…but there’s still a problem” he then began as she glanced to him. “Reports have been coming into me since I came about an outbreak in Sector A…troops have been saying the people we left behind are rioting…getting to the point they’re taking out our own,” he said while she stood. “Shall we deal with it?” he then asked.

Thinking to herself for a moment, she soon looked back towards the city before she simply shook her head and spoke.

“No, I have no time for useless people. Handle it,” she said firmly to him as he slowly nodded his head looking at her.

“Of course, Commander” he said, walking away while she continued to keep her eyes ahead.

As she did from afar, Jimmy and Adrian could be seen towards the east side of the city as they set up their camp across a bridge. With a river flowing underneath it as it was an entrance point into the city, Danny could be seen looking through a pair of binoculars as he stood on the rooftop of a building. Looking on, he soon took out his radio as he saw explosions beginning to be set off ahead.

“We got movement ahead…looks like it’s starting,” he said looking down as Adrian nodded her head to Jimmy.

“Let’s move out then,” she said as she turned away to head out with the other soldiers. As they did, Jimmy soon pulled out his radio as he began to speak.

“Lutheran…you copy?” he asked as Lutheran spoke. “Yeah,” he said panting while Jimmy spoke.

“Are you in position?” he asked while Lutheran shook, running through the streets while HOPE soldiers followed behind him.

“Almost…we’re heading there now,” he said as Jimmy nodded and continued to follow Adrian.

While the HOPE soldiers stood miles away from the scene, Emily kept her eyes ahead where in the distance, she too could see the explosions beginning to erupt within the city. Standing still as pillars of smoke rose up into the sky. She simply smiled while IOBs could be seen screeching and moving through the streets. Hitting the traps she had placed, she soon looked at a soldier beside her and spoke.

“Send in the first unit,” she said as the soldier nodded.

Soon troops in her forces began to move out. Heading through the city, soldiers quickly ran while the blood of Emily ran through them. Running quickly as their speed increased, they soon spotted the IOBs coming ahead and they stopped and began firing.

Shooting at the IOBs, the creatures quickly went down, screeching and running forward while the troops faced them.

With them clashing, gunshots could be heard echoing across the city as Emily stood. Watching her soldiers strike, she then sent out another unit of soldiers to help.

Heading out, the soldiers carried a gun that sparked at the end of their barrels. Eventually heading towards the scene, the second unit of troops stood firing out bullets that shot out and tasered the IOBs. Piercing their skin, the bullets begin to shock the IOBs while they went trembling down.

With the creatures distracted, the soldiers soon began to push forward.

All around the IOBs screeched while they were shot back by the forces of Emily, who viciously advanced ahead.

While they did, Emily stood in place. Looking up into the sky, she noticed a figure in the sky. Appearing before her, the IOB warship revealed itself from under its cloak.

Looking up ahead, Jimmy and Adrian watched with the other troops as they too could see the massive metallic ship as they traveled through the city.

Laying itself still within the air, Emily then watched as troops below it began firing towards it. With bullets and explosives deflecting all off against the metal, they soon watched while the smoke cleared and the ship began to open up.

Slowly above it, a port opened up then while a device could be seen levitating up. Heading up into the sky, it eventually stopped while it scanned the city below. Holding itself at bay for a moment, it soon began to activate as it released a blue, web-like substance, forming across the city.

Slowly developing over her, Emily smiled while the troops around her talked amongst each other as they watched the web.

Forming over the entire city, the device eventually placed a shield in place so no one could enter it or leave. Trapping everything below it, Jimmy looked on, breathing heavily as he began to get nervous. Trying to stay focused Adrian soon looked at him as she quickly shook her head and spoke.

"C'mon Jimmy! Keep going!" she said as he quickly looked back at her and nodded as he kept his eyes ahead.

With the shield fully up around them, Emily then looked at the soldier beside her and spoke.

"So they finally show themselves..." she said as she shook. "Let's see how menacing they are now...all units engage them at once!" she said as the soldiers nodded.

Vastly more soldiers began heading out, in units of hundreds as they went towards the scene. Running all out before her, she looked at the ship hovering in the air as her soldiers went to aid the others.

Quickly, holding out their guns before them, they began firing at the IOBs that were being blown back and screeching across the streets.

While more of her forces entered the scene, the IOB induced soldiers quickly went up to the IOBs grabbing them by the head and smashing them to the ground. Other soldiers could be seen running into them and picking them up while the creatures retaliated. Soon soldiers were having blades and claws pierced through their bodies while some were blasted by the IOB's mouth cannon.

Even broken and shattered, the soldier's bodies quickly continued to heal as they continued to fight against the IOBs.

Seeing her forces prove themselves far superior than the IOB creatures themselves, she soon looked up towards the warship in the sky as she soon saw it begin to make its first move.

With a port opening up below it, it soon dropped a device that slowly levitated towards the ground. Slowly falling, soldiers below could be seen fighting. It soon began to light up as from the ship, a massive red cannon shot down onto the device and it went crashing into the ground.

Smashing into the city, the device let off a massive shock wave. All around, the concrete grounds were being torn with the wave as buildings and streets were being torn into pieces.

Quickly, IOBs and soldiers were being destroyed within a 5 mile radius as the shockwave rattled the city. All around it could be seen. Jimmy and Adrian with the rest of the HOPE forces continued to advance while they stumbled, moving through the city. Even with Emily, she could feel the building rattle and quake while she stood on its top looking at the ship in the sky.

With the shockwave eventually passing, the city could be seen with only dirt and broken pieces of concrete underneath the ship. Wiped out all underneath it, the cannon underneath the warship steamed while it stood above the city still.

Scattered across the city, soldiers of Emily could be seen lying on the ground as they struggled to move. With limbs missing, they began to slowly heal and form their parts back while the IOB warship soon made its next move.

With soldiers trying to recover all around underneath it, it then began opening small ports underneath as large metal pods were being released. Stretching out across the city, the metal pods went crashing down into the concrete and they soon began clanking open.

Releasing a mist of warm steam, the sounds of screeches could be heard as groups of IOBs were released. Crawling out of the grounds as they appeared in numbers, they quickly went running out through the city as they began finishing the soldiers trying to stand.

Even away from the distance, Emily's troops could be seen advancing ahead as the scout ships in the sky appeared over them and continued to fire down at them.

Seeing soldiers being wiped out all around her, Emily still stood while she then looked to the building beside her.

Coming into view, it was there where Kate could be seen below. Standing in place, she saw Emily nod to her as she smiled and began walking ahead. Quickly taking off her gear and walking towards the edge of the building, she soon stretched out her arms while she dropped off it.

With the air hitting her face, she began falling down the building, her eyes soon went black as she then changed into her flying IOB form.

Letting out a piercing screech then, she then brought herself up and began flying over the street ground as she headed towards the battle.

With her massive wings flapping in the air, she continued screeching while she flew ahead.

Seeing soldiers struggling within the streets, she quickly then let out a roaring screech and released a stream of fire to towards the incoming IOBs. Screeching as they looked up, she engulfed them in a blanket of flames while more of Emily's forces were sent out. Pushing ahead then as they followed Kate's lead, the soldiers ran through the fire and began clashing against the army of IOBs.

All around the sounds of fighting continued while the soldiers tried to strike down as many IOBs as they could. Piercing knives and bullets into the creatures, the soldier's eyes glowed black while the blood within them was causing them to slowly change. Becoming more vicious, the soldiers could be seen fighting with their bare hands as they even clawed through the IOB creatures.

Gaining the upper hand though, Emily stood smiling watching the lines of smoke appearing in the streets. With Kate still gliding across the city as she lit the IOBs on fire, the scout ships in the city quickly noticed her.

They all immediately stopped zigzagging across the city as they flew towards her. Soon beginning to pursue her, they began firing towards her, while the shots simply deflected off her skin. Screeching as they moved all round her, she soon lunged out at one catching it in her mouth and it soon exploded.

Setting off in the air, she soon flew through the smoke as the ships still gave chase.

With the ships distracted, more of Emily's troops began to move in. Moving in small taskforces, they quickly scattered to areas in the city where troops needed support.

Striking down IOBs as they began to surge a push, Emily nodded as everything was going to plan. Watching the war continue, she then watched before her, as a soldier yelled.

Looking ahead, an IOB scout ship could be seen getting blown up by a soldier on a rooftop. Ignited on fire, the ship continued to crash towards her while the soldiers beside her began to back up.

Seeing the ship coming, Emily simply stood still when all of a sudden, a rocket could be seen firing towards the ship and it exploded before her.

Blowing up, soldiers all around the building fell back while they yelled. With the ship stopped, she stood watching as the smoke cleared and the pieces of the ship started falling to the ground.

Thinking for a moment, she soon glanced behind her while the soldiers slowly got up to their feet.

"It seems we have some visitors," she then said as she looked back down to the ground. "Find out who let out that shot and kill them," she said as she soldiers behind her nodded.

They moved out while she looked back towards the war going on beneath her. Lutheran could be seen standing below in the street with the other HOPE soldiers around him. Glancing at her as he tried to stay hidden, he simply shook his head as he looked at the soldiers with rocket launchers beside him.

“It’s no use…take to the roofs. We’ll attack her from there,” he said quietly while they all nodded and began to move out.

Chapter 32

Jumping from a building, Randall could be seen crashing down to the street as other soldiers followed. Slowly rising up, he cracked his neck as he looked behind him to see the war beginning. Smiling, he soon looked ahead of him as he began walking the streets of Sector A.

Looking around, he walked slowly.

The scene was quiet. No sounds or any kind of movement could be seen.

Humming to himself while he walked, he eventually made it to a body of an IOBCF soldier lying on the street. Nodding his head as he looked at it, he looked up and around him as people of Sector A could be seen on the rooftops. Seeing them beginning to surround him, he soon nodded his head as he spoke.

"Well, well, well…what do we have here?" he said looking around as he could see the people of sector A pointing guns down at him. "I can't say I'm surprised. You people sure don't know when to give up," he then said as he looked at the older people and even children holding out guns towards him. Shaking, while they held their ground, Randall simply chuckled as he spoke.

"Alright, let's settle this before anyone gets hurts now, shall we?" he then said as he looked around. "Who is the one in charge here?" he then asked as it went quiet.

With only the sounds of the war going on behind him, he simply stood as he held out a gun and shot it into the air. Letting off a few shots, the people of Sector A nervously kept still while Randall spoke again.

"Let's try that again before the next one goes in someone's head," he then said. "Who's in charge?" he asked as all of a sudden, a female voice appeared from behind him.

"I am," she began as Randall looked behind him to see Liz standing outside the window sill of the building. Looking at her, he simply squinted his eyes while he shook his head and spoke.

"Wait…but…" he then said as all of a sudden on top of a building, Holly appeared, landing down in her IOB form. Seeing her there, his eyes soon flew wide open, while she then lifted up her massive head and let out a piercing screech.

Hearing her screech, he simply knelt down slowly grabbing his ears while the other troops fled. Retreating as they left him there, he could only simply watch as he looked back towards Holly and was slowly engulfed in a stream of fire.

Chapter 33

With the war continuing, IOBs quickly climbed on top of the buildings and began firing blasts down to the streets. Shooting down below, soldiers in the streets were fighting and soon blown back from the blasts. With their skin burning and bodies trying to heal, the soldiers kept on trying to fight as they could only watch IOBs run up to them and tear them apart.

With more pods being released from the IOB warship, Emily stood still while a soldier beside her soon spoke.

"We're ready for the next phase, Commander," he said as she then nodded. "Sound the alarm then," she said as the soldier nodded.

Eventually then, an alarm began to sound off across the city. The same alarm system in Sector A, troops fighting in the streets, kicked back the IOBs while they turned and heard the call. With it sounding off, the soldiers quickly began looking at each other and yelling out to retreat while the IOBs quickly followed. Screeching and trying to reach them, the soldiers led the IOBs away from the IOB warship where they then stopped. Turning back to the creatures, the soldiers watched as the herds of IOBs running to them soon hit mines that went off across the city. Blowing up with smoke rising up into the air, Kate soon swung over the soldiers and engulfed the street full of IOBs on fire.

Lighting them up, the creatures were blocked by a wall of flames while troops on the building tops fired down below.

Shooting down at the IOBs below, a soldier soon stood firing when all of a sudden, he looked back as Danny knocked him out. With him falling, the other IOBCF troops looked to him as they too were soon knocked out from behind by other HOPE forces.

With soldiers being taken out, the fire in the street soon began to disperse while Emily's forces ran forward.

While they did, they soon watched as cans of gas were soon thrown out in front of them.

Releasing a thick cloud of smoke, the soldiers began coughing and stepping back while HOPE soldiers quickly ran out to the scene. Quickly showing themselves, the HOPE soldiers fired tranquilizers into Emily's soldiers who eventually collapsed down in the street. Cutting them off, the line of HOPE soldiers traveled inward to keep pushing back Emily's forces, while another group began firing towards the IOBs ahead.

Seeing the troops pouring out into the scene, Emily then stood looking toward them. Seeing them split her army away, she watched Jimmy appear at the scene. Looking at her, he soon put a gas mask on and turned away with Adrian while they went to attack the IOBs.

With them pouring into the streets, Emily merely shook her head as she turned to the soldiers behind her and spoke.

"Take out the solider in white…kill them all," she grunted when all of a sudden, the soldiers looked behind her and she turned to see rows of missiles fire at her.

Blowing up before her this time, the building top began to collapse while the building itself fell. With soldiers around her being blown back, debris and smoke rose up while HOPE soldiers hit the other building tops as they began to take out the IOBCFs.

With the building collapsed, Emily could be seen standing while smoke and debris lay all around her. With bodies of the soldiers lying in the concrete behind her, she stood with her eyes closed as she soon looked ahead. Seeing through the smoke as HOPE soldiers head towards her, she soon grunted while she slowly turned into her IOB form.

Transforming, the troops slowly walked towards her. Not able to see her within the smoke, she soon came out screeching while she quickly slashed her blade across a soldier and pierced it through another. Pulling the blade out of the soldier, she turned to the others who were firing at her.

With them shooting into her skin, she quickly lunged at them while she grabbed the head of one and threw him to another while she then pierced her hand through another soldier. Hearing him grunt while she held him up, she soon threw him aside while she stood looking at the dead soldiers lying around.

Standing, Lutheran quickly lunged out behind her as he jumped onto her back. Screeching as he bit and held onto her neck, she furiously shook him off and tossed him away as he fell to the ground rolling.

Quickly getting back to his feet, he stood lowering his head and growling while Emily viciously turned to him and spoke.

"You again," she grunted while he shook his head and spoke. "It seems you have knack of getting in my way…I find it hard to believe the last time we met, you got the better of me…pretty impressive coming from a mere dog," she said as Lutheran spoke.

"I should have killed you when I had the chance," he snarled as Emily shook with laughter.

"Haha don't be a fool," she then began. "If you've come to kill me…then you truly are as stupid as you look," she said as he spoke.

"We'll soon find out then, won't we?" he said as she shook.

"If that's the case…then I'll show you how dangerous I really am, dog," she said as Lutheran simply growled and ran towards her while they soon collided.

Chapter 34

While Lutheran began his fight with Emily, Jimmy stood in the streets as he fired towards the IOBs. Screeching and running up to him, while he and Adrian shot them down, he soon looked up as he noticed a scout ship passing by. Seeing it come towards him, he quickly ducked down while he yelled out to the soldiers.

"Take cover!" he said as he and Adrian ducked down while the ship fired its lasers into the street. Firing into the ground while debris fell all around them, Jimmy stood back up with Adrian as they continued to fire ahead.

Fighting off the creatures slowly with the other HOPE soldiers shooting out more cans of gas and tranquilizers into Emily's forces, Kate could be seen flying over the city as she continued to light the streets on fire. With more scout ships trying to stop her from behind, the IOB warship soon turned its attention to her while it's cannon moved. Aiming towards her in the air, it soon began to charge up while it let off a massive shot that darted off across the sky.

With Jimmy looking up with the other troops, the blast shot out beside Kate as she screeched and felt a massive hit. With a shockwave dispersing in the sky, while the ground shook again, Kate soon went twirling down from the sky and into the city. Crashing down, she soon collided with buildings and debris that scattered all around her as she found herself changing back to her normal self while she lay on the ground.

Clay could be seen sitting at the ledge of the building next to Mya. With their feet dangling off the edge, Clay simply looked down, thinking to himself, while Mya soon began to speak to him.

"You've been in a lot of pain, Clay. You've been hurting yourself," she said as Clay slowly nodded while he thought.

"After you left…it's been complicated." he then began as she nodded. "Things haven't been making sense and I didn't know what to do," he said as she spoke.

"When you realized that I was gone," she then began as he sat. "Your mind pushed me away…it realized that I wasn't real and because of that I couldn't exist anymore," she said as he shook.

"I…I never meant for this to happen," he said as she shook.

"I know Clay…and I didn't want to leave…but I couldn't come back until your mind was clear," she said and Clay nodded as he looked away.

"And that's why Jake brought me here now…" he then said. "To get you back," he said as she shook. "Not just that…but to help you realize who you are again, Clay," she said as Clay shook. "You've given up, Clay, you've let yourself fall so far that you don't see who you are," she said as Clay shook. "Why Clay? Why are you doing this to yourself?" she then asked as he shook.

"Because I don't have a choice, Mya…there is no choice that was given to me…and I can't do anything but give up," he said as she shook.

"Clay. You can't…you can't give up" she said.

"I'm sorry, Mya…but I just can't take this anymore. I can't take being afraid. Being afraid of who I am and what I stand for," he stated. "I don't want anyone else to get hurt because of me. I don't want to see anyone in pain when I'm the one that should take it," he said as she shook.

"But Clay, people in the world need someone like you. Someone who stands up for them and can believe in…" she said. "They want to be strong like you, to step out into the world and not be afraid, to be free and stand up for themselves," she stated. "They have a right to feel this way, but they need someone who will be with them. To show them the way," she said as Clay shook his head.

"Even if I could, Mya. I don't see who could possibly need someone like me," he said. "Clay, there are people in the city right now who are looking for someone like you," she said as he stood.

"People who are being led by Emily, right now and without you, she is going to work them and kill them to the point where there is nothing left of them," she said as he thought. "If you don't stop people like Emily, they will continue to crush every bit of hope in people in the world until they are exactly the same and that's exactly what Emily is trying to do. She will bring every single one of those people to their knees like slaves and kill them until they all feel that same pain she feels," she stated. "And the sad thing is, she's only the beginning Clay there are far greater threats out in this world that you've overlooked," she stated. "Threats that are causing pain to people each and every day," she said as he then shook.

"What…what do you mean?" he asked as Mya shook.

"Clay…there are things out in the world far worse than Emily. Things that are truly horrible than they have ever been," she said as he thought. "There are people who are falling victim to cannibalism, rape and abuse…people who are dying of hunger and sicknesses…people who are dying of wars and violence…people dying from being sold in slavery…all of these things are still happening in the world…things that are happening now. And the worse thing is that all of these are happening on top of the IOBs inhabiting this planet. IOBs who are still killing, torturing, experimenting and abusing the world too," she said as he shook. "You see, Clay. You can't give up," she said as he turned to her. "Just because you're afraid, doesn't mean you have to quit. People are relying on you…people who are looking for hope during these times. Someone who will stand up and give them a voice in places that may seem impossible. They want to be strong and be able to live a life without fear, even with IOBs or even with people. You can bring that out in them. And people may get hurt, people around you will get hurt, but if they see you, they will know that their sacrifices are not in vain, that everything they have been working for, won't go to waste when they see you because they know you got their back and Jimmy was right when he spoke to you, the world needs someone like Clay Treston, even if he' doesn't. Because if there's anything I know about you, Clay…it's your will to help and inspire people and I know you'll do the right thing," she said as he stood thinking as she spoke. "This is so much bigger than any of us, Clay, so much more bigger than you and deciding whether or not to fight, means everything in the world to someone who is trying to survive now," she stated. "You are that hope for people and if you give up now, then you are giving up on these people," she said as he slowly thought shaking his head while he spoke.

"I…I don't know if I can do this alone though," he said as she shook.

"And you won't be," she said taking his hand. "You don't have to be," she stated as he looked to her. "No matter what…I will always be by your side and no matter what, I'll always make it back to you…but I need you to come back to me now. To do what is right and what I need you to do," she said, standing before him. "Do what you always promised me, don't let the world share our pain. End it," she said kissing him on the cheek as he slowly closed his eyes.

With his feelings coming back to him, he soon found himself waking up. With his eyes wide open, he looked all around as he noticed he was on the floor. Breathing heavily, Jake soon walked up to him and held out his hand as he spoke.

"Well…I have to admit, I wasn't sure if you would make it back. Kind of glad to see that it worked," he said as he helped Clay up.

"What…what happened? What did you do to me?" he asked as Jake spoke.

"Well, to be honest, I gave you something that potentially could have killed you, but you were strong enough to come back," he said.

"So I almost died?" he asked as Jake nodded.

"In a way, yes but I was hoping it would help you realize a near-death experience, a moment where you decided whether to give up or not, and I take it you had something that led you back," he said as Clay thought looking at the ground while he looked to see Mya standing beside him again. Standing still as he believed in her again, he soon spoke.

"You could say that," he said looking at Mya as Jake then patted him on the back.

"Well, it's good to see you come back to life again," he said walking around. "For a moment, I thought I had really lost you," he said as Clay shook.

"Yeah well, I may have found my way back but that doesn't change what's happening in the city," he began looking at a monitor in the front of the room, as Emily was already in the city. "She's risking hundreds or even thousands of lives out there…if I don't think of something quick, I won't be able to stop her," he said.

"Maybe…but if you remember, I may have something that might be able to stop her," he said as Clay looked at him while he thought.

"Will it work?" he then asked as Jake spoke.

"Well, it depends, Clay," he then began as Clay stood. "It depends if you're willing to do what it takes to do it," he said as Clay shook.

"But I-" he began as Jake quickly lifted up his hand.

"Just because you're back…doesn't mean you're ready," he said as Clay shook his head in confusion. "You need to know what's at stake here, Clay. The people that are relying on you are not just here in the city," he began. "There are people in the world that need your help, a world on the verge of completely losing itself. If you fail here, it won't necessarily mean you lost, but it will determine your fate from here on out." he said as Clay stood looking at him. "You have this ability in you, Clay, to defend people from ever experiencing pain, ever feeling pain you've felt and…if you don't answer those cries, then no one is going to see the world and life the way it really is, understood?" he spoke.

"Yeah…but what does that have to do with anything?" he asked as Jake chuckled.

"What I'm trying to say is, Clay. There is a path written here for you. A path that will lead to great things if you manage to succeed. The question is, are you ready for it? Are you ready to put aside yourself and your self-doubt, your pain, your past, and to accept things as they are now? Move on to actually do something great in this world? Something that is not supposed to be done but something that has to be done?" he asked, while Clay stood thinking to himself. With feelings running all across his body, he soon looked up at Mya walking up to him as she slowly put her hand on his face and nodded. With her next to him, he soon looked back at Jake slowly nodding, while he stood, with a war already beginning.

Chapter 35

With IOBs heading towards him, Jimmy quickly lifted up his gun as he shot down the creature. Screeching down next to him, other HOPE soldiers came up beside him as they fired at the incoming IOBs. Holding their ground as they continued to push back the IOBs, an IOB on a building soon let off a blast that exploded before Jimmy.

Getting shot back, the other soldiers were soon struck down by IOBs as Jimmy struggled to get up. Leaning his head up, he watched as an IOB stood over him and chomped its jaws at him. Holding it by the neck, he struggled to hold the screeching creature back while it clawed at the ground beside him and drooled on his face.

With its teeth chomping closer to him as he struggled to keep it off, Adrian soon came up beside it as she quickly struck a knife in its side.

Screeching as she did, it soon stood up, turning to her as she quickly struck the gun across its face. Falling down beside Jimmy, she then quickly turned around and fired at the rest of the IOBs. They quickly fell, riddled with bullets.

Taking them down, she soon stood up, turning to Jimmy as she then helped him back up to his feet while she spoke.

"Hey, you ok?" she asked as he slowly nodded.

"Yeah I'm good," he said as all of a sudden, they watched the IOB Kate fly over them as she went up ahead lighting the streets on fire with IOBs burning within it.

The fire holding the creatures back for a moment, Jimmy shook his head while he continued to look around.

"Well, that will buy us some time," he said while they stood hearing calls behind them. Turning back, HOPE soldiers could be seen walking backwards while they fired towards Emily's soldiers who were breaking through the ranks and attacking.

Seeing the soldiers come back, Adrian soon took out a hand gun and flipped it down beside her while it swung into a blade. Having it near her, she began marching forward as Jimmy quickly held onto her arm.

"Wait! What are you doing?!" he asked as she shook in his grip.

"We don't have a choice! We need to split up even more to hold off Emily's forces!" she said.

"Yeah, I know…but the whole point is to not kill them!" he said while she looked to her blade. Hearing more soldiers cry out as they were being overwhelmed, she turned towards them thinking. She soon looked back to Jimmy.

"Then I'll improvise!" she said as he nodded. "Just hold off those IOBs!" she said running away, while Jimmy looked behind him to see the flames in the street ceasing.

With the flames going out before him, he watched IOBs stepping out of the flames, letting out screeches as he could only shake his head to try to halt their advance.

Lutheran could be seen lunging out towards Emily as he caught hold of her arm. With her stepping backwards as he held on, she quickly then lifted him up over her head and threw him down to the ground behind her. Hitting the ground hard, he let go of her as he watched her bladed hand come down. Rolling away, the blade went striking into the concrete, while Emily soon watched Lutheran fire out tranquilizers into her.

Attaching to her skin, she simply wiped them off while she let out a screech and fired blasts out of her mouth.

Firing all-out before him, Lutheran quickly moved and rolled across the street while the blasts shot all around him.

Moving quickly, while he evaded the shots, he soon made it back to his feet while Emily quickly then lunged forward into him. Rolling across the street, she eventually pinned him down to the ground while she bit her jaws into his armor. Trying to bite through, she couldn't pierce through his armored vest as he quickly then deployed a gun beside him. Turning towards her quickly, it began firing rapidly into her face, while she let him go.

Stepping back, Lutheran went back down to all fours as he deployed another gun beside him. With him still firing bullets into her, she continued to cover her face, while she crouched down. Holding on until his bullets eventually ceased, she furiously stood back up. With the bullets falling out of her skin while she healed, she quickly opened up her mouth and let out another blast that shot out before him.

Shooting out towards him, he quickly moved away, while the energy of the blast pushed him away.

Rolling across the street, he soon looked back at Emily who kept lunging forward and they continued to fight.

Chapter 36

Colliding together, HOPE soldiers began fighting off IOBCF forces that were eventually forming up around them. Firing and shooting down the soldiers, IOBs quickly appeared attacking both while they screeched, tearing into the soldiers. With an all-out war breaking out in the streets, the HOPE soldiers looked up to see scout ships firing towards the buildings they stood on while they collapsed. With soldiers being blasted away, HOPE troops soon found themselves retreating from the scene while the IOBCF troops ran ahead to continue their fight with the IOBs.

Trying to push them back, the IOBCF's threw aside their weapons as they found their teeth sharpening. Losing control, they quickly began drooling by the mouth while their nails grew. Grunting and snarling, they quickly ran towards the creatures, where both found themselves tearing into each other.

While the troops dispersed, Kate could be seen waking back up where she lay in the debris. Looking up, she noticed the IOBCF troops still fighting while the scout ships continued to fly overhead. Seeing them flying, she soon quickly stood back up and changed back to her IOB form to give chase.

Heading into the skies, she quickly grabbed hold of one, exploding it in her claws while she waved her tail across another. Screeching as they crashed all around, she soon let out a stream of fire towards them to hold them off.

With the fighting continuing, Clay stood as he watched the top of the room lift up while he could see the grey skies above him. With ash falling down, he simply looked around him as Jake spoke.

"I managed to steal this ship some time ago…I figured it would come in handy one of these days," he began as he walked up to Clay. "Only thing is I have to keep it hidden all the time or else people will find me. Not really fond of visitors, you could say," he said as Clay nodded.

"So where do I go from here?" he asked.

"We're already in the city. I managed to get us here before the shield went up. We're not far from where the action is. Traveling by foot will allow you to avoid the unnecessary battling going on right now. Your focus is on Emily and Emily alone…so that's where you should go," he said as Clay nodded.

"And what do I do when I get to her?" he asked as Jake chuckled and pulled out a small tube-like container and threw it to Clay and Clay caught it.

"It's a serum," he said as Clay pressed a button on it and it deployed out a little stub needle from the end of it. "Once you see an opening on Emily, directly put the serum into her chest, from there, the serum will destroy the IOB cells within her body and change her back to her normal self again. But you only have one chance, Clay, so use it wisely…anywhere else the serum is placed, may not reach her heart in time to spread the serum around in her…so be careful," he said as Clay looked to him.

"And what about the rest of the people she's changed?" he said as Jake shook his head.

"I'll explain that afterwards, but for now, focus on just Emily. The rest of the people will be fine," he said as Clay nodded. He looked to the serum in his hand as he clasped it in his palms and he looked back up at Jake.

"Jake. Thank you…for everything," he said.

"Don't thank me yet, not until this is all over…until then, remember this isn't about anyone else. It's about you. Protect yourself, Clay, and good luck," he said as Clay nodded and he jumped down from Jake's IOB ship. He looked out to see it lift back up in the air. As he did, he heard Jake's voice talking through the collar on his armor as he spoke.

"By the way, I put a communication device in your suit. I'll do what I can to have eyes on you throughout the battle. Hopefully, I can help you out," he said as Clay nodded and looked towards the city before him.

Chapter 37

With soldiers clashing, Kate looked ahead and she could see the IOB warship in her sight. Letting out a screech while she headed towards it, scout ships appeared from behind as they tried to shoot her down. The lasers deflected off her skin, she headed for the ship that formed a shield around it. Kate quickly lunged into it while the shield shook. Screeching while she was halted in the air, she quickly let out a stream of fire over it as the ship's shield held.

Breathing fire over it, she then flew up in the air as the scout ships following her ran through the air while they went crashing into the shield. Blowing up and sending an electrical wave through the shield, it dispersed in the air as a shockwave burst through the air. Waving across, Kate quickly went towards the ship and landed on it. Ripping her claws through it, she began screeching and tearing her claws in the metal of the ship while pieces went flying off. Trying to hold her off, the ship began firing dispersing missiles into the air, they simply flew past her and shot all across the city. The ship was unable to stop her in time and she quickly bit her jaws into it and tore off the top. Ripping it off, the IOBs within all stood looking up, shooting blasts from their mouths up at her as she let out a breath of fire over them.

Covering the entire inside of the ship in fire, it exploded from within and a massive explosion went off in the air. Smoke and debris shot out all across the sky, Kate came out of it screeching and looking at it crash down into the city. While it did, the device over the city began to flicker and it soon dispersed. Dropping down from the sky, the shield around the city began to clear away with the ship gone.

As the ship fell, Adrian and Jimmy looked up at the sky to see the shield dispersing across the city. Watching it disappear, Adrian looked at the sky to see pieces of debris falling down, heading for Jimmy. Seeing them coming, she quickly struck down an IOB-induced soldier, as she quickly sprinted towards him. Looking around up at the sky, Jimmy watched paralyzed as he too could see the debris heading towards him. With his eyes wide open, he could only feel himself stand still. Adrian quickly ran into him, pushing him away at the last second while the debris landed in the street.

Chapter 38

Seeing their warships gone, the scout ships around the city began scattering, while the IOBs on the ground slowly stepped back, seeing their ships leaving.

With IOBs beginning to retreat from the scene, Emily's forces continued to push forward while they pursued the creatures escaping.

Lying on top of the building, Danny continued firing tranquilizers into Emily's troops while they ran by. With so many appearing in the streets, he made it back to his feet and picked up his gun and began ordering the HOPE forces to retreat from the scene.

While they retreated, Lutheran snarled towards Emily as she swiped him away. Falling to the ground, he rolled away as she came running down with her blade while he tried to move away. Hitting the concrete, she stood back up as he quickly jumped towards her back again, biting into her neck. She quickly moved around, trying to get him off as she then reached back with her claw.

Scraping her nails over his face, he soon let go as he fell back to the ground. Getting to his feet, he shook his head with the marks across his face as he looked back at her with one eye. With him trying to stand and fight, she only continued lunging forward. He was forced to continue and fight.

Chapter 39

Slowly getting up, Adrian pushed a piece of concrete off her leg. Grunting as it fell back, she breathed heavily the dust from the debris floated in the air. Trying to get up, she noticed she couldn't stand as she simply sat looking around. With a massive piece of the IOB warship on fire behind her, she looked at Jimmy lying on the ground, not far away.

"Jimmy…" she then said as she tried to go to him.

Eventually, dragging herself over to him, she sat down next to him. With him unconscious, she lifted up his head and put it on her lap, she tapped his face and spoke.

"Jimmy…Jimmy, wake up…wake up!" she shouted as he wasn't waking up.

Feeling his pulse as he was still alive, she could only sit wondering, what to do next as the sounds of war surrounded her.

Thinking, she looked ahead and noticed something, pieces of concrete moving. Looking at them, she watched as an IOB crawled out and stood in the street.

Snarling because of its wounds, it simply stood before her as her breath trembled. Reaching for her leg, she noticed her gun wasn't there, as she quickly looked at it. Wondering where it went, she looked around and found it was lying on the ground, where she had lain.

Unable to reach it, she looked back at the IOB ahead.

Looking around as it moved, it soon turned as its massive eyes locked onto her.

Seeing her sitting, it began chomping its jaws. With its claw moving, Adrian could only sit there shaking her head as she spoke.

"No…no please…" she said to herself while the creature soon lifted up its head and let out a screech. Screeching, it then began running towards her as she sat, unable to get away.

Sitting still, she could only lower her head and close her eyes while she held onto Jimmy.

With time slowing down, she sat shaking while she waited for the IOB to come. Waiting to be struck, she began remembering her brother. Images started flashing in her head, she remembered moments being with him as kids. From playing in the parks, to being at home. She specifically began to imagine him as she remembered her final moments with him and the note he gave her. All rushing to her at once, she felt someone touch her shoulder as she all of a sudden heard her brother's voice.

"Adrian…" he said as she slowly then opened up her eyes.

Running towards her, the IOB soon came over to her. Raising up its blade hand, it came down towards her but she quickly grabbed it.

Grunting as its eyes went wide, Adrian could be seen holding the IOB's blade in her hand while she bled. With blood running down her arm, she soon looked up towards the IOB before her as she spoke.

"I won't let you kill me…not here! Not now!" she grunted as her eyes went black. Changing before the IOB, her hand soon formed into a blade as she stood up and struck it through the IOB.

Holding it in as the creature screeched, she soon had her eyes change back, while she let it go. Falling back as it did, she stood breathing heavily while she looked at her bladed hand. Seeing it, it soon changed back to normal as she stood thinking about what happened.

Trying to figure it out, she heard Jimmy waking up behind her as she turned to him and knelt. "Jimmy! Jimmy, are you alright?!" she began as he slowly lifted himself up and nodded.

"Uhm…yeah…I'm fine," he said as he then looked at her hand. "You're bleeding…are you alright?" he asked as she nodded.

"I'll be alright," she then said as she looked back beside her. "C'mon…we need to get to Lutheran," she said grabbing her gun, as she then helped him back up to leave the area.

Chapter 40

Firing bullets rapidly, Lutheran stood watching as Emily furiously screeched towards him. He was unable to stop her and she lunged into him as he fell to his side. Sliding against the concrete, she soon stood back up.

Grabbing him by the head, she lifted him up.

Smiling while she held him, she then threw him hard to the ground and his body bounced off the ground.

Whining as he hit it, she soon picked him up again and threw him towards a rusted old car. Smashing into it, the glass shattered all around him as he fell to the ground.

Struggling to get up, he could only look at her as she walked towards him. Trying to stand, he soon found himself fall because he couldn't.

With no other choice, he could only watch Emily slowly head towards him.

Seeing him down, HOPE soldiers appeared on the scene as Danny looked to see Lutheran lying on the ground. He began shooting towards Emily with the other soldiers as she stopped and turned towards them.

With the bullets not affecting her, she let out a blast towards them.

With it blowing up near him, Danny covered his face while the soldiers were blown back beside him.

Protecting himself, debris shot all around him as he quickly looked back towards her and saw a blast heading towards him.

Seeing it coming, he quickly ran to his side but the explosion blew him back, off his feet.

Rolling across the ground, Danny lay unconscious while Emily stood smiling and looking back towards Lutheran.

He was still trying to get up. She came up beside him and smashed her foot down onto his face. Pressing his face against the concrete, she stood looking at him as she then spoke.

“It’s only a matter of time now dog…” she began moving her feet away from his face. “With the IOBs gone, you and rest of these HOPE soldiers will be nothing but dust when I’m through with you,” she said as he tried to get up. “Tell me something though…why do you even side yourself with these humans when you have no place in this world with them?” she asked as he grunted, moving his head still being pressed to the ground while he looked at her.

“Because…no matter how much I change, I’ll always be one of them,” he said glaring at her while she shook.

“Tsk…then you are truly pathetic aren’t you dog?” she began stepping off of him while she knelt down and put her blade on his neck. “It’s a shame…I think I really could have used you,” she said while he lay there slowly growling. Chuckling as she looked down at him, she watched as a blade shot down beside her.

Striking the concrete where she stood, she recognized the blade, and turned around to look behind her. Staring in the direction of where it had been shot from, she then saw none other than Clay standing in the distance.

Chapter 41

With the blade planted next to her, Emily stood looking at Clay who slowly lowered his arm on the building rooftop. Standing still while he looked at her, she lunged into the air. Clay backed away, waiting for her.

Furiously jumping into the air, she went crashing into the buildings while she clawed her way up. Eventually pulling herself up, she landed on the building Clay stood on. Slowly breathing, she began standing back up as she looked towards Clay and spoke.

"It's you…ha…it's really you," she began as she stood, changing back to her normal form. With her standing before him, Clay simply stood still looking toward her as she was back in her human form.

Shaking her head then, she spoke, "I have to admit Clay…I didn't think you would be still alive." She then said, walking around, "I considered you being different…but not to the point you can defy death" she said.

"All that matters is I'm here…that should be your only concern," he said, looking at her as she chuckled.

"Well regardless…I'm glad you came. I was beginning to get bored with your friends here…they sure can be a pain in the ass," she stated.

"This isn't about them, Emily…it's about you and me…it always has been since the start…and it's time to end it," he said.

"Haha really, Clay?" she began. "You really think you can come here and stop me? You think all of this will be undone by defeating me?" she asked as Clay stood still, the ashes slowly falling down around him.

"If that's what I have to do to end this…then I'm willing to do whatever it takes," he said while she thought.

"You're a fool, Clay…you're just as blind as the rest of them," she said as he shook.

“Maybe…but I know that it didn’t have to end this way, Emily…it didn’t have to end in us fighting,” he said.

“Look around you, Clay” she then said. “This isn’t a fight…it’s war,” she stated. “A war that I’ve been fighting for years…a war where the only way to get my freedom, is to destroy the freedom in others. I worked and sacrificed everything to get to this point. I destroyed everything I cared about to survive and I’m not going to let you stop me now,” she stated.

“If that’s what you want…then fine,” he said cringing at her. “But no matter what I do, I have to stop you Emily…” he said as she quickly smiled and her eyes turning black.

“Then it’s time I get rid of you for good and show you who I really am!” she said grunting while she all of a sudden began to transform back into her IOB form. With her body changing before him, Clay felt a warm breeze pass around him and Mya could be seen walking towards him. Soon standing by his side, she shook her head as he stood looking toward Emily.

“This is it, Clay,” she then stated as he nodded. “Put away the fear in you right now…think about the people that need you. Use it and decide who you are…” she said as he nodded, spreading his feet apart and deploying his blades. Holding them out beside him, he slowly stood ready while Mya continued to speak. “Make it known and show her, Clay. Show her here and now what you can do!” she said. Emily drooled from her mouth and let out a screech while she ran towards Clay.

With her running towards him, Clay stood waiting for her to come as she quickly lifted up her blade. Throwing it down, Clay lifted up his arm as she slid the blade across his gauntlet.

Holding it up, he quickly spun gliding the blade across her stomach as he then spun back looking at her. Seeing her wound slowly heal, she lunged towards Clay screeching, while he backed away from her bladed hand. Swinging it all around, he moved to his side as he threw the blade down onto her arm. It fell to the ground. But as Clay looked at Emily’s arm, it began to form back while she threw her blade forward into his chest.

Hitting his armor, Clay fell straight back while he rolled up. Standing back on his feet, he watched as her arm formed back while she continued to screech forward.

Flying over the city, Kate screeched over Emily’s forces, continuing to chase away the IOBs. Driving them away, the soldiers below slowly began to change even more while their heads began growing. With their teeth growing more and their arms becoming longer, they were turning more and more into IOBs with the blood still running through them.

As they began to change even more, Kate soon flew in the air as she looked back towards Emily. Seeing her begin her fight with Clay in the distance, she let out a screech and began to head towards them.

Fighting all across the building top then, Clay could be seen swinging the blade around as he tried to face Emily. With him evading the swings of her blade and claws, he moved around quickly, trying to look for an opening with the serum.

Eventually, she caught her claws across his arm as he yelled. Quickly kneeling down as he held the claw, she kicked him back.

Falling down onto his back, she stood over him and threw her blade down onto him. Seeing her blade descending, he quickly crossed the blades before him as her blade clanked against his. With the blade nearly touching his face, he grunted, trying to hold her back as he looked at her before him.

Refusing to give up, he moved the blade, pointing towards her shoulder and shot it into her. Screeching as she came off of him and stepped back, Clay soon got back to his feet with another blade loading as he lunged towards her.

Pulling out the blade and throwing it, Emily looked back at him as he quickly came up to her, gliding a blade across her arm. Catching her as it fell off and disappeared, he kept swinging the blade at her.

Moving around, he furiously pushed forward while she screeched and caught her blade across his cheek.

Catching him, she let out a blast from her mouth at him. Seeing it coming, he crossed his arms before him as the blast caught him in his armor.

Getting blown back, he landed on the ground still covering himself, while his arms steamed.

Simply staying where he lay, Clay opened his eyes back up as he breathed heavily. Hurting, his body ached, he continued to try and get up. With her wounds healing and her arm reforming, Emily then chuckled before while she spoke.

"Well c'mon Clay! You're going to have to do better than that if you want to beat me!" she grunted as Clay breathed slowly, trying to catch his breath while Jake spoke to him.

"Clay…it's working!" he said as Clay stood. "Her healing abilities are slowing down. But if you don't find your opening for the serum soon, it'll be too late," he said as Clay looked at her wounds, now completely healed.

"I…I don't see it though" he said.

"Don't worry, you will…the more Emily is wounded, the more she'll have to work to heal, eventually, she should slow down enough for you to use it," he said as Clay looked and saw Mya beside him. Looking at her, he nodded as Emily simply walked up to him and lunged at him with a screech as they continued to fight.

Chapter 42

While Clay continued to fight, Adrian and Jimmy headed towards the scene. Running, they noticed IOBCF soldier standing before them and they quickly began firing. Ducking down, they hid behind the debris, Adrian quickly put her back against the wall structures and began loading her handgun.

Turning and firing towards the soldiers, Jimmy looked around, thinking out aloud.

"We'll need to find another way around…c'mon," he then said, cutting through an alley as Adrian looked at him and slowly began to follow.

Clanking blades together, Emily swung her blade over Clay as he knelt down. When he rose back up, she quickly threw her bladed hand forward while Clay crossed his arms. Hitting the armor on his forearms, he stepped back while he looked to her. Breathing heavily as he tried to think of what to do, he looked beside him to see Kate in her IOB form flying next to him and the building.

Seeing her, he quickly rolled away while she let out a stream of fire towards him.

With Kate missing him, he slowly sat on one knee as he watched her let out a piercing screech while she flew over to Emily.

She landed down on the building and it shook. Clay slowly stood back up as he looked at the both of them. With Kate standing still behind her, Emily then spoke.

"I have to admit…you're more skilled than I thought," she said.

Clay looked to them thinking as he spoke, "You know, it's funny." He caught his breath. "All that strength…and yet it takes two of you to kill me," he said as the IOB Kate drooled from her mouth while Emily stood next to her.

“It’s simple, Clay…” she chuckled. “Having us both here ensures that you die this time,” she said as Clay simply shook his head. Mya walked up beside him.

“Clay…you can do this. Stay focused,” she said as he breathed heavily. “You’re worried…you don’t have to be. She’s afraid. That’s why they’re both here. She knows she can’t win alone,” she said as Clay spoke.

“Even if that’s the case…” he said heavily. “I can’t beat them both,” he said as Mya shook, holding onto his hand.

“Regardless, we’ll do this together,” she said as he looked at her and nodded, before he turned back towards Emily and Kate, who stood before him.

Chapter 43

While Clay and Emily fought, Lutheran was looking toward them and he could see Clay was in trouble. Breathing heavily, he stumbled around walking while his body was trying to heal. Eventually falling to the ground, he began growling to himself as he forced himself up. Meanwhile Clay found himself facing Kate as she rose up.

With more fire streaming out to him, he quickly rolled to his side. Rolling onto one knee, he watched Emily lunge out at him.

Swinging her blade over him, he knelt down as he glided his blade across her leg.

Standing back to his feet, she quickly stumbled while Kate looked down at Clay. Looking toward her, she let out another stream of fire as he ducked down trying to block the fire.

Burning all across his arm as he yelled out, he soon fell to his side. Trying to get up while the side of his body was burned, Emily appeared beside him as she threw down her blade.

Holding his arm out as she clanked against his armor, he struggled, holding her back. Pushing down on him as she drooled, he soon kicked her back, while she backed away enough for him to stand.

As he stood though, he then watched as the IOB Kate by his side threw her massive tail and it hit Clay and sent him across the building top. Stumbling as he rolled around, he merely rolled off the edge of the building as he slipped his blades into the concrete. Stopping himself from falling, he looked over the edge behind him while Emily and Kate looked at him.

Eventually, getting back to his feet, he almost fell as the burns across his skin continued to ache. Kneeling down, he held himself up with a blade as he slowly looked at Emily before him.

"You should have never come back here, Clay," she began as she toggled around. "You should have never come back here," she began as Clay looked at her. "Your city is lost…I destroyed it and I've taken control of these people. Why bother fighting for them when they don't believe in you anymore? Why fight for a city that's lost?" she asked as Clay breathed heavily and he shook his head and spoke.

"Forcing people to kill other people…while convincing them to fight for you," he said catching his breath as he spoke, "that will only make things worse." Emily shook her head as she spoke, "The world is already a mess, Clay…you can't change that…as long as IOBs live here on earth, this world will never have a chance." "If anything I'm helping them…I'm making them stronger and it's only a matter of time before they take back their world and I will rule for my own! Where no one will dare cross me!" she said.

Shaking his head as he breathed heavily, Mya then kneeled down beside him and as he then spoke. "It's too much, Mya," he began. "I can't do it…I can't beat them…not like this," he said as Emily simply looked at him, chuckling.

"No, you can do this, Clay…you need to stay focused," she began. "Outsmart them…do something they won't expect!" she said, as Clay nodded trying to think of what to do.

"You truly have lost your mind, haven't you, Clay?" she then said to him. "You keep dreaming about a world where people can be different…a world where you think they can hide to survive…but in reality, deep down, even you know that I can save these people, you know that in the end, I can change the world," she stated as Clay shook while he slowly looked up at her.

"No, that's where you're wrong," he said as he reached for a device beside him and clicked onto his blade while he pointed it towards her. Firing, the blade shot out past her as she saw the blade land in the concrete before Kate.

Seeing him miss, she simply looked back at him and spoke.

"Haha really, Clay? Is that it?" she said, as Clay quickly then pulled out a device in his hand with his thumb on a button. Her eyes quickly then widened as he spoke.

"No…not yet!" he said as he pressed it. The blade had an explosive attached to it and it went off, blowing up the entire top of the building as Emily screeched and Kate flew off. The blast shot off into the sky as the building below it began tumbling down.

While it did, Jake from his ship soon stood up from his chair slowly, while he watched to see if Clay would make it.

As the building collapsed, the city ground rocked. Jimmy, from afar, stopped as he looked up with Adrian at the building collapsing under a mist of debris and smoke. Seeing it come down, he quickly went on ahead with Adrian to find their way to Lutheran.

While they did, Kate flew in place above the collapsed building, looking down at it, as smoke hung around blocking her view. Trying to see if anyone survived, she continued to fly around the scene while pieces in the debris began to move.

Coughing as he shoved the pieces off, Clay appeared. He was slow to get up as he was on all fours. Crawling out of the debris as the concrete fell off him, he breathed heavily trying to catch his breath and his vision was blurry. With marks across his armor and skin as he tried to see through the smoke, he simply looked up and he heard Kate still trying to investigate the scene as he watched her fly over.

Seeing her above, Clay made it to his feet slowly as Mya walked up beside him and nodded. As she did, Jake sat back in his chair while he spoke looking at him.

"It's not over yet, Clay, she's still there, keep an eye out," he simply said as Clay stood. Jimmy arrived at the scene and as he stood looking around he noticed Clay in the distance.

Seeing him standing, Jimmy quickly chuckled as he shouted out towards him, "Hey! Clay!"

Clay glanced back at him calling out as he spoke. "Jimmy…" he began in confusion as he turned to him while Jimmy stood waving.

But as he did, Jimmy's eyes soon grew wide open while he looked towards Clay. Seeing the expression on Jimmy's face, Clay slowly turned around and grunted while he felt a blade pierce through his body. Feeling the blade in him, he slowly then looked up to see Emily standing behind him as she chuckled.

She then raised Clay up in the air with the blade still in him as Jimmy quickly yelled out and ran towards them. But as he did, he was soon cut off by a line of fire that shot out from above, blocking his path. Falling back on the ground as he looked up above the flames. Kate flew over him while Adrian quickly grabbed him to pull him back.

"No! Wait, I have to help him!" he cried out as Adrian grabbed him back and spoke.

"Then get up and let's find another way!" she said holding out her hand as he quickly nodded and grabbed it and they went to find another path towards Clay.

As Jimmy did, Clay was still being held up in the air as Emily chuckled, holding him up while she then slammed him on the ground. Grunting and yelling out as he was thrown down, Emily pulled out the blade from him slowly as he lay on the ground, bleeding.

Smiling, Emily then grabbed him by the head with her claws as he bled out and slowly then began dragging him with her.

Eventually throwing him up against a broken concrete wall, she pinned him against it as she held him up.

Struggling as she held him by the throat, he stuck out his hand telling her to stop, while she then shoved it back and pierced the blade through it. He yelled as she pinned the blade in his hand to the wall, she simply looked to him bleeding out while she slowly spoke.

"It hurt, doesn't it, Clay?" she said as he struggled to move while Kate landed behind her in the street. "Did you really think you could win, Clay? You really thought you had a chance?" she said as Clay shook his head while he was pinned. Trying to speak, he could only look behind Emily where Mya stood. Looking at her face as she watched him struggle, he looked back at Emily, while he spoke.

"Even if I fail…at least, I know that I tried…at least, I would know…that I did everything I could," he said as Mya looked at him and her eyes watered as he looked at her with a smile coming onto his face. "Even if I die…then at least, I know…I would finally be somewhere…somewhere where I can be free…away from the pain…away from the world…and be with the one I care nothing more about…" he said as Mya looked at him smiling back. Emily shook her head as she tightened up her grip around him as she spoke.

"You go after nothing more than a dream…" she began. "If you wanted that…then why are you still here?" she asked as Clay shook his head and he looked up to the sky.

"Because…I realized…I still have more I need to do, before I see her…" he said as Emily chuckled while she held him.

"Haha you sound as if you haven't failed yet," she began. "What more could you possibly do?" she said as she squeezed his throat while he struggled. Barely holding on, he simply shook his head then while he lifted up his other hand. Looking at him lift it up, she soon noticed a device in his hand as he held his thumb over the button and spoke.

"Beating you…" he said.

All of a sudden, Emily stood hearing a device going off. Beeping behind her, she soon looked back while she looked at Kate standing and all of a sudden, the device could be seen underneath her. Immediately as they saw it, the device went off. Blowing up beneath Kate, a massive blast shot up into the sky while Jimmy and Adrian looked up at the explosion going off. Seeing it as it shook the ground, Jimmy could only shake his head while he spoke.

"C'mon! We need to hurry!" he said while she nodded. They went to find a way through the debris. Emily covered her face, as she was forced to let go of Clay on the debris. Holding onto his wound, he looked forward to see pieces of the creature Kate falling down from the sky.

As the smoke cleared, Emily slowly looked around to see her Kate gone and she firmly clenched her teeth together. Furiously, looking back at Clay, she then shook her head.

"You! You fool! Do you have any idea what you've done!" she said as he looked to her. "You'll pay! You'll pay for this!" she said holding her blade hand back when all of a sudden, beside her, Lutheran lunged forward growling as he grabbed hold of her arm and she backed away, trying to shake him off as he held on. As Lutheran attacked, Clay grunted in pain as he forced himself up and Mya and Jake spoke.

"Now, Clay!" they said as he stood up while he watched Emily toss Lutheran off her. Eventually tossing him, Clay quickly pushed himself forward as he yelled raising a blade while cutting off her bladed hand from behind.

Screeching, she looked as she realized her arm was gone. She turned to him and grabbed him as he looked and saw her arm not growing back as quickly. Seeing it, Emily held him by throat as beside them, Jimmy moved in firing at her. She quickly dropped Clay. Looking at him and Adrian, Adrian soon fired a grenade that blew up onto Emily as she fell back.

Landing at the ground, Clay immediately stood back up and he went and quickly ran almost on top of her as he came down with his blade in her chest, and she grunted and pushed him off beside her.

As Clay rolled away, Emily then leaned up as Adrian shot another grenade towards her as Emily covered her face with one arm as it was blown off.

Laying on the ground as she had no arms while they tried to regenerate, she soon tried getting up when all of a sudden, she watched as a stream of fire came down around her. She quickly got to her feet, looking up as before her, Holly came down, letting out a roaring piercing screech. Liz appeared standing on her back as she held out her gun and pointed it at Emily.

"You're still alive?!" she said as Liz simply shook her head as she spoke.

"It'll take a lot more than that to kill us, you bitch," she said while Holly then screeched and let out a stream of fire towards her. Covered in fire, Emily fell back to the ground hard as she rolled across the concrete.

Scrambling around as she tried to get up, Lutheran then appeared again as he jumped up on her and bit into the back of her neck as he tried holding her down from behind.

Screeching, she soon grew back half of her other arm as she elbowed him off and Lutheran fell back.

As Lutheran fell, she got on one knee and behind her Clay quickly went up to her once more and he shoved the blade into her back. She lunged forward, barely catching herself from falling back.

Holding herself up, she shoved her head back, hitting Clay and he pulled out the blade from behind her. Stepping back, he watched her arms finally regenerating. With them forming back, Clay quickly shook his head and furiously ran once more into her, putting his shoulder to her back as she fell forward. Landing in the debris, Clay quickly held her down and immediately put a blade down into her back as she clawed at the ground, struggling to get away.

Elbowed off her one more time, Clay had to back away while she stood back up, turning to him. As she did, he lunged forward, piercing the blades through her again while she fell back. Falling, Clay held a blade into her chest that went down into the street concrete beneath her. Screeching, she shoved her blade hand into his shoulder as he grunted, with her claws scraping across his face and head. Holding her down, he quickly then glided the blade across her hand as it fell. As it hit the ground, Clay then quickly waved the blade and swung it down towards her neck while he held it still. Feeling the metal touch her, while she was pinned down, she only could look at Clay breathing heavily over her. With sweat running down his face, he slowly then shook his head as he looked down at her and spoke.

"It's over, Emily…" he said as Emily shook and looked to him. "You can still change…you can still stop all of this!" he said as she shook her head and smiled as she looked at him.

"The only way to stop me is to kill me here, Clay," she said as he looked down at her. "If you want me to stop…you're gonna have to cut off my head," she growled as Clay shook his head as he looked at her.

With everyone standing behind him, Clay slowly stood back up and looked down at her. Reaching for the serum on his belt, he soon shook his head as he slowly spoke.

"That's where you're wrong," he said as he quickly came down closer to her chest. She grunted while Clay held it in his hand.

Injecting the serum into her chest, Clay watched as she lowered her arm and slowly began changing. Getting off her as he took out the blade in her chest, Emily slowly turned back to normal while her wounds healed and her arms transformed back to be human once more.

"What…what did you do to me?" she then asked looking at herself become normal again. Seeing her change, Clay simply stood looking at her as he shook.

"Your IOB form…it's gone…and so is your war," he said as she furiously looked at her hands and shook while she looked at Clay.

"No! No! You can't! You can't do this!" she yelled, standing back up.

Looking down at Clay and the others, looking at her, she simply thought to herself while she spoke. "No…no, I don't want this! I don't want a life like this again!" she said as she then looked below her and found a piece of glass. "I won't let you do this to me…I won't!" she said as she then grabbed it while Clay's eyes opened up.

"No Emily! Wait!" he said running up to her as she smiled and put the shard of glass through her chest. Piercing the glass through her heart, she pulled it out. Looking at her, Clay watched as she slowly collapsed as he ran up to her. Quickly going up to her, Clay knelt down as he quickly caught her before she touched the ground. Holding her up, he gently then set her down on the ground. Setting her head down on the debris, he quickly turned around while he yelled at Liz behind him.

"Liz! Hurry!" he said as he put his hand over her wound. Trying to cover it up, she slowly grabbed hold of his wrist as she shook her head, breathing heavily.

"No…no, more pain…no more" she said as Liz stopped behind them. "I don't want it…I don't want it taking over again," she said as Clay shook his head while he looked at her.

"No…No, it doesn't have to end this way. You have to hang on," he said as she soon chuckled while she lay on the ground, thinking.

"I'm…surprised, Clay, after all I've done, yet…you…you actually cared about me," she said while Clay looked at her quietly.

Seeing her look away, she found herself looking up towards the sky while she felt her body going cold. With nothing but the view above her, she soon had a tear run down her face as Clay sat watching and she eventually went still.

Chapter 44

Slowly, a steady snow of ashes continued to fall down upon the broken city of Hollandview. With it going silent, it wasn't until hours later, as the evening skies came down onto the city. With HOPE soldiers gathering around, exhausted and in pain, that Adrian walked among them while they took Emily's body away.

Standing, while she let out a deep breath, she then looked beside her to see Jimmy walking up to her as he slowly spoke.

"Hey," Jimmy began as Adrian nodded. Standing next to her, they watched the soldiers walking around the street as he eventually spoke. "You ok?" he asked as Adrian shook.

"I guess I don't know how to feel right now," she said as she shook her head. "All this time…All I wanted to do was stop someone from hurting other people…to do something I would be proud of….but in the end, after watching Emily go, I question what I really was fighting for," she said while he thought. "I watched a woman…a woman who used to be like us, be so tortured by the world around her that she took her own life. I don't think I can ever say that I made my father proud for doing this," she stated while Jimmy shook.

"I think in a way, we're all to blame for what happened here, Adrian," he stated while they stood thinking. "But you did what you could…And I think that's something your father would be proud of," he said as she shook.

"Maybe…but I wish I knew…I wish I knew he would have been proud of me," she said as they stood. Thinking for a moment, Jimmy slowly then pulled out an envelope in his jacket as he held it out to Adrian while he spoke.

"Here," he said as she shook.

"What…what's this?" she began.

"Something I should have gave you a lot sooner," he said as he then handed her the envelope that had her name written on it.

Opening it up, her eyes began looking around the paper while she spoke.

"It's a letter…from my father," she said as Jimmy nodded.

"He wrote it before he knew you were in the camp," he said as she then chuckled, looking at the letter with her eyes watering. "Regardless of what you did…he was always proud of you, Adrian…" he said while she spoke putting her hand over her mouth.

"Thank you, Jim…" She quickly then went up and hugged him. Standing there, holding her, he simply then shook his head.

"If anything, Adrian…it should be us thanking you," he said while she held him tightly.

As she did though, she looked back at her hand to see the cut from the IOB fully healed. Looking toward it as she thought, she continued to hug Jimmy while they hung in the city below.

While they did, Clay could be seen from afar. Standing on the ledge of a building, he looked around at the city before him. Watching over it, he stood talking with Jake on the other end of his ear piece.

"With Emily gone…all that's left is to cure the people she changed," he began as Jake nodded within his ship. "You think they'll be ok?" he asked.

"Haha, well, I would be surprised if they weren't," he stated as Clay thought.

"So in the end?" he asked as Jake chuckled.

"Let's just say eventually, because people don't always produce IOB cells within them, their abilities should eventually cease once the cells from Emily die out," he said as Clay nodded and looked ahead while Jake spoke. "So to answer your question, they should be fine Clay….You did good," he said while Clay shook.

"Well, not exactly…" he then said. "I had no intention of Emily getting hurt like this…and instead, she's gone," he said while Jake thought. "I was hoping I could convince her to change maybe…Convince her that there was more to fight for in this world…Just like what you did with me," he said .

"Well perhaps this happening…shows another reason why people need our help, Clay…" he said as Clay nodded and thought. Looking down then, he could see Liz and Holly walking through the streets together, helping wounded soldiers. Standing beside each other again, they looked at one another, happy, while Clay looked back.

"I take it you won't be sticking around," Clay then began as Jake nodded within his ship. "Will I see you again?"

"There's a good chance," Jake then replied. "I wish to someday find a new home elsewhere but not until I undo a few other things here on Earth…until then, if you ever need help, you can guarantee I'll be there," he said as Clay nodded.

"Well, until then," he said. Jake sitting back in his ship, smiled. "Good bye, my friend," he said as he then turned off the ear piece.

Hearing him gone, Clay slowly took off the piece while he held it in his hand. Standing once more by himself, Lutheran then appeared behind. Walking up to him, he slowly sat down beside Clay, looking toward the city. Together with Clay, he then shook his head as he thought and spoke.

"It's strange…seeing a city free like this again," he began with Clay nodding. "It's almost hard to believe," he said.

"Hopefully, it won't suffer anymore. Suffer from anyone like Emily or the IOBs again," he said as Lutheran shook as he spoke. "To be a city of hope for once and not a city of death,"

"Well as long as IOBs are alive, the suffering will always seem to continue," he said while Clay thought. Thinking to himself, Lutheran then cleared his throat as he looked at Clay beside him and spoke.

"Look, Clay. I'm sorry for what I said earlier…you could say I wasn't myself, maybe too. I uh, I didn't look through the pain you were going through…let alone did I see what I really meant to say to you," he began as Clay looked down, thinking. "To see you come back though anyways to help us…makes me only feel selfish to talk about you like I did…especially with Mya…I wouldn't lie when it comes to me saying this, if anything, she would be proud of you" he said as he saw Clay smile as he looked down at Lutheran beside him and spoke.

"Even if you hated me, Lutheran, you will always be like a father to me," he said to him. "Ever since I've met you…you've done nothing but look out for me…and you've given me advice which at times I don't always see right away…but you're always there…and quite honestly, if you didn't tell me what you did, I don't know if I would have been here today," he said while

Lutheran nodded. "Besides…I think I still owe you for saving my life again," he said as Lutheran could only chuckle as he soon shook his head and spoke.

"Even so…you won't have to worry about me. At least for a while…" he said as Clay thought for a moment.

"Your son…you think he's still out there?" he asked as Lutheran nodded.

"Maybe. But for now, it's just about taking care of some loose ends," he said looking at Clay. "Things I need to do before I join you and Jimmy," he said while Clay nodded his head and he spoke.

"I know what it's like…looking for answers," he began as Lutheran nodded. "It takes time…but eventually, you find them," he said looking at Lutheran as he spoke. "You do what you have to do for now, Lath…and if you need my help, I'll do whatever I can," he said as Lutheran looked at him and he nodded and sat looking back at the city.

"The HOPE forces are going be lucky, having us with them," he then said as Clay smiled. "Hopefully we'll make a difference with them then what we managed to do here" he said.

"Hopefully…" Clay said, taking a deep breath. "But no matter what, we'll always keep moving forward now…it's all we can do," he said.

"It'll be a long journey," Lutheran said as he stood. "Life outside of here will probably be no different…full of pain and war…war against people…war against IOBs…no matter what we do, I'm sure one threat will come after the other," he said as Clay nodded. "I guess the question is…are you ready for it?" he then stated as Clay looked ahead at the sun that began to shine through the clouds once again.

Looking at it, he felt the warm air pressing against him that he had felt so long ago. With a new world before him, he stood looking at a city of his past. A city of suffering; And before him a new journey lay. A journey that will lead him to not only a new beginning. But a new life.

Thinking, he soon shook his head as he spoke.

"Regardless," he began as he looked beside him. Seeing Mya stand next to him, he slowly let a smile play on his lips as he slowly took her hand. Feeling himself alive once again as she stood beside him, he slowly spoke while he looked into her eyes, "I am now."